COLORIMETRY AND IMAGE PROCESSING

Edited by **Carlos M. Travieso-González**

Colorimetry and Image Processing
http://dx.doi.org/10.5772/intechopen.68246
Edited by Carlos M. Travieso-González

Contributors

Humberto Moreira Villegas, Leticia Álvaro, Anna Melnikova, Julio Lillo, Jianjun Chen, Abdullah Beyaz, Baraka Maiseli, René Vázquez-Jiménez, Raúl Romero-Calcerrada, Rocío N. Ramos-Bernal, Patricia Arrogante-Funes, Carlos J. Novillo, Sulpicio Sánchez-Tizapa, Ali Zaher, Chiman Kwan, Jin Zhou, Bence Budavari, Jesús Raúl Martinez, Ernesto Martinez, Miguel Martinez, Manuel Miranda, Humberto Cervantes, Eduardo Quevedo, Gustavo Marrero Callico, Hiroshi Kamada, Tomohisa Ishikawa, Keitaro Yoshikawa

Published by InTech
Janeza Trdine 9, 51000 Rijeka, Croatia

Publishing Process Manager Dajana Pemac
Technical Editor SPi Global
Cover InTech Design team

First published January, 2018
Printed in Croatia
Legal deposit, Croatia: National and University Library in Zagreb

Additional hard copies can be obtained from orders@intechopen.com

Colorimetry and Image Processing, Edited by Carlos M. Travieso-González
p. cm.
Print ISBN 978-953-51-3744-3
Online ISBN 978-953-51-3745-0

For EU product safety concerns:
IN TECH d.o.o., Prolaz Marije Krucifikse Kozulić 3, 51000 Rijeka, Croatia,
info@intechopen.com or visit our website at intechopen.com.

Meet the editor

Carlos M. Travieso-González received his MSc degree in 1997 in Telecommunications Engineering from Polytechnic University of Catalonia (UPC), Spain, and his PhD degree in 2002 from the University of Las Palmas de Gran Canaria (ULPGC, Spain). He has been an associate professor since 2001 in ULPGC, teaching subjects on signal processing and learning theory, and he got his positive accreditation from the Spanish government to be a full professor. His research lines are biometrics, biomedical signals and images, data mining, classification system, signal and image processing, machine learning, and environmental intelligence. He has researched in more than 45 international and Spanish research projects, serving as head researcher to some of them. He is the coauthor of 4 books, coeditor of 19 proceeding books, and guest editor for 6 JCR-ISI international journals and up to 20 book chapters. He has over 375 papers published in international journals and conferences (58 of them indexed on JCR, ISI, and Web of Science). He has published four patents, and three more are under revision in Spanish Patent and Trademark Office. He has been the supervisor on 7 PhD theses (8 more in process) and 120 master's theses. He is the founder of the IEEE IWOBI conference series and president of its Steering Committee and founder of the InnoEducaTIC conference series. He has been the reviewer in different indexed international journals (<58) and conferences (<180) since 2001.

Contents

Preface

The origin of images and videos has been a revolution for society, and different technological advances have shown their different characteristics, in both professional and social settings. In particular, the use of smartphones and multimedia contents has increased their applicability. Since their origin, the quality of images and videos continues to improve.

In this book, various applications for the improvement and use of color images and image processing will be described, which can be applied to provide more information regarding their graphic contents.

In the fields of security, medicine, and communications, new and advanced techniques can be applied to facilitate or give more information about multimedia contents. A great number of tools are being developed in this sense, and this book presents high-quality work, developed with scientific methodology, and giving validation to the present proposals. The work focuses on colorimetry and image processing and hopefully will be a very informative read.

Colorimetry and Image Processing is composed of 10 chapters, which have been divided in two sections: colorimetry and image processing. The section "Colorimetry" has five chapters. Colorimetry is observed by using: a visual interactive programming learning system, an approach based on the color analysis of the habanero chili pepper, an approach based on scene image segmentation centered on mathematical morphology, other systems based on the simulations of dichromatic color appearance, and finally, an approach based on color reconstruction to enhance the use of super-resolution methods. The section "Image Processing" also contains five chapters. Image processing is shown by using: pansharpening algorithms for hyperspectral images, an approach based on the analysis of low-resolution satellite images and a ground-based sky camera for estimating cloud motion, a hybrid super-resolution framework that combines desirable features of TV and PM models, a study of real-time video analysis used for anthropometric measurements on agricultural tools and machines, and finally, an approach based on the threshold optimization iterative algorithm using ground truth data and assessing the accuracy of a range of threshold values through the corresponding kappa coefficient of concordance.

As editor of this book, I would like to thank the authors for their effort and dedication in achieving a work of great quality. The sum of this effort has produced this book, which has become an must-have read for all those who want to know the latest advances in colorimetry and image processing.

Carlos M. Travieso-González
University of Las Palmas de Gran Canaria,
Spain

Section 1

Colorimetry

Chapter 1

Colorimetry and Dichromatic Vision

Humberto Moreira, Leticia Álvaro,
Anna Melnikova and Julio Lillo

Additional information is available at the end of the chapter

http://dx.doi.org/10.5772/intechopen.71563

Abstract

Normal trichromats have three types of cone photoreceptors: L, M, and S cones (most sensitive to long, medium, or short wavelengths, respectively). Therefore, standard colorimetry is based on three variables (X, Y, Z). Dichromats only have two types of functional cones due to genetic factors. The main consequences are that dichromats (1) confuse colors that can only be discriminated by the response of the type of cone they lack and (2) make errors when naming colors. Chromaticity diagrams can be used to specify dichromats' color confusions. Confusion points represent imaginary stimuli that only activate L, M, or S cones. Confusion lines radiate from confusion points and represent pseudoisochromatic stimuli (i.e., colors confused by the corresponding type of dichromat if presented at an appropriate intensity). Dichromat's color appearance models have been developed to simulate the colors supposedly seen by dichromats, and there exist color simulation tools that implement some of those models.

Keywords: dichromacy, color vision, color simulation, color naming, color preference

1. Introduction

Normal trichromats have three types of cone photoreceptors in the retina: L, M, and S cones (most sensitive to long, medium, or short wavelengths, respectively). Therefore, standard colorimetry, which was developed for normal trichromats, is based on three variables (X, Y, Z) [1], and it is not directly suitable for severe color vision deficiencies like dichromacy [2].

Dichromats only have two types of functional cones due to genetic factors [2, 3]. Protanopes, deuteranopes, and tritanopes lack L, M, or S cones, respectively. Protanopes and deuteranopes are also known as red-green dichromats and constitute the most frequent form of dichromacy, affecting about 2% of human males. Tritanopia is a much rarer disorder that affects less than 0.01% of humans. The structure of this chapter reflects the main consequences of this physiological feature for dichromats at different levels: (1) color discrimination, (2) color appearance, (3) color naming, and (4) color preference.

Regarding color discrimination, dichromats confuse colors that can only be discriminated by the response of the type of cone they lack. How can we use standard colorimetry to characterize dichromatic color vision? Chromaticity diagrams can be used to specify dichromats' color confusions if we assume that dichromacy is a reduced version of trichromacy [4, 5]. In brief, confusion points represent imaginary stimuli that only activate L, M, or S cones. Confusion lines radiate from confusion points and represent pseudoisochromatic stimuli (i.e., colors confused by the corresponding type of dichromat if presented at the appropriate intensity). Section 1.1 will describes the rationale of this procedure in greater detail and will introduces a feature present in most part of red-green dichromats, the so-called residual red-green discrimination, which to some extent can challenge the validity of this procedure.

Although chromaticity diagrams are a powerful tool to characterize color confusions in dichromats, they do not tell us anything about the appearance of colors in this kind of observers. Dichromat's color appearance models have been developed to simulate the colors supposedly seen by dichromats, and there exist color simulation tools that implement some of those models. Section 1.2 will describe in some detail one of the most famous color appearance models in dichromats [6].

1.1. Chromaticity diagrams, confusion lines, and cone fundamentals

1.1.1. Chromaticity diagrams and confusion lines

Under the assumption that dichromacy is a reduced form of trichromacy (*reduction, loss,* or *König hypothesis*), it is possible to use standard chromaticity diagrams in order to characterize dichromatic color confusions [4, 5]. **Figure 1** shows confusion lines radiating from confusion points for protanopes (first row), deuteranopes (second row), and tritanopes (third row) in *CIE 1931 xy* (left panel) and *CIE 1976 u′v′* (right panel) chromaticity diagrams. Confusion or copunctal points represent imaginary stimuli that only activate a given type of cone and can be determined by the intersection of several confusion lines for a given type of dichromat. For example, the protanope confusion point, determined by the intersection of several protanope confusion lines, represents an imaginary stimulus that only activates L cones (protocones), without triggering any response in M cones (deuteracones) or S cones (tritacones).

Confusion lines or pseudoisochromatic lines represent stimuli that are indistinguishable for a given type of dichromat (if presented at the appropriate intensity). For example, red-green dichromats (protanopes and deuteranopes) accept the full range of mixtures of a monochromatic light of 545 nm (green) and a monochromatic light of 670 nm (red) to match a reference monochromatic light of 589 nm (orangish yellow) of adjustable intensity. As it can be seen in **Figure 1**, spectral lights of 545, 589 and 670 nm fall in protanope and deuteranope confusion lines. This match is known as Rayleigh match, and it is implemented in the Nagel anomaloscope, the most precise and reliable instruments designed to diagnose color vision deficiencies. Pseudoisochromatic plates, widely used to detect color vision deficiencies, use different combinations of surface color stimuli selected from different confusion lines (for a full description of color vision diagnose, see [7]).

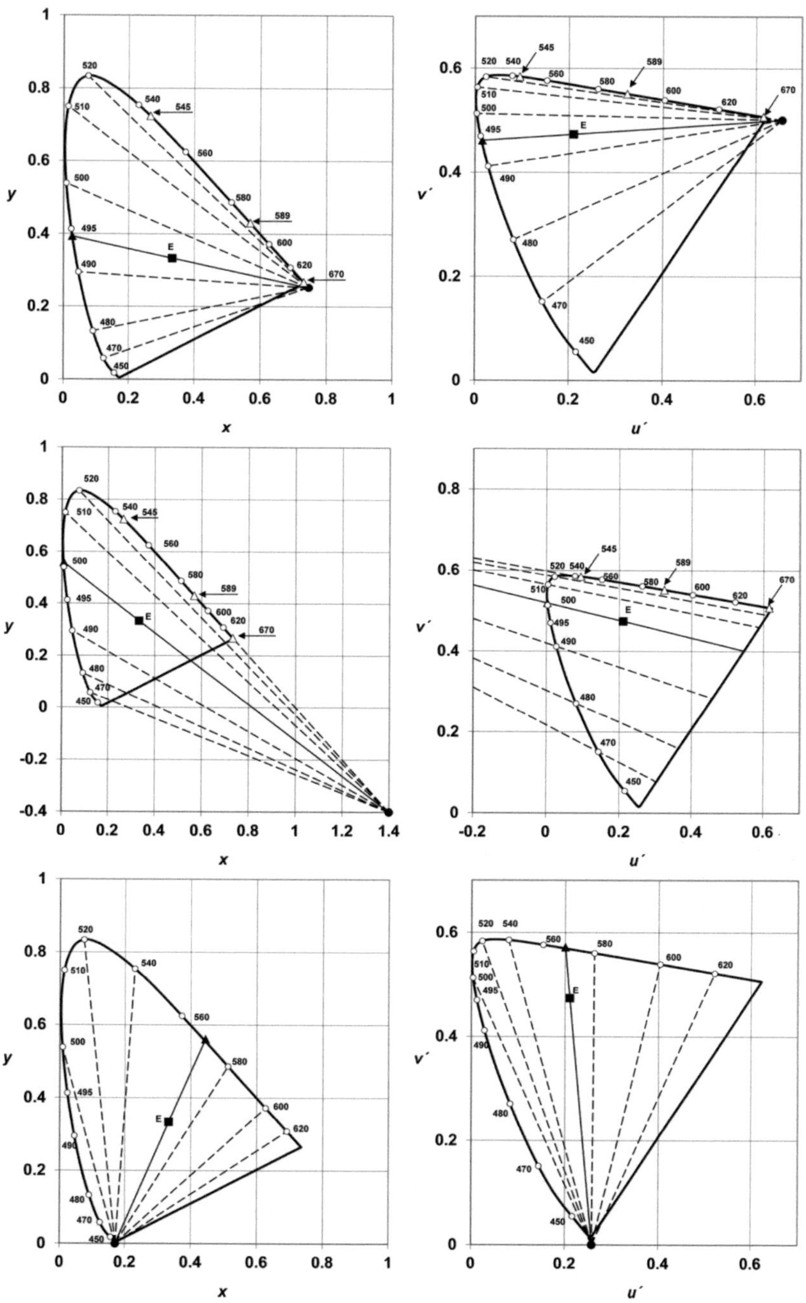

Figure 1. Confusion lines and types of dichromatism. Dichromatic confusion lines represented for protanopes (first row), deuteranopes (second row), and tritanopes (third row) in CIE *xy* 1931 (left panel) and CIE *u'v'* 1976 (right panel) chromaticity diagrams for different wavelengths (dashed lines) and the equal-energy stimulus E, represented by a black square (solid line). Black circles represent protanope (x_p = 0.747, y_p = 0.253), deuteranope (x_d = 1.40, y_d = − 0.40), and tritanope (x_t = 0:171, y_t = 0) confusion points. Black triangles represent the neutral points for dichromats (see Section 2.2). The stimuli of the Rayleigh match (white triangles) are also represented for protanopes and deuteranopes (see text for details).

Nevertheless, the use of pseudoisochromatic lines to characterize the color confusions of dichromats is limited mainly due to the following factors: (1) the existence of individual differences in relation to the spectral sensitivities of the cones in a given group of dichromats [2, 3] and (2) the existence of *residual red-green discrimination* in the major part of red-green dichromats for color stimuli over 3° of visual angle, revealed by the fact that red-green dichromats behave as anomalous trichromats for large-field stimuli ([8, 9]), as well as the existence of *residual S cone function* in some observers diagnosed as tritanopes [3].

1.1.2. Cone fundamentals derived from color matching functions and dichromatic confusion points

Cone primaries or cone fundamentals (i.e., the spectral sensitivity of the cones) derived from confusion color matching functions (CMFs) and dichromatic confusion points are known as *König fundamentals* and have been extensively pursued in color science (the interested reader can consult a large list of references, e.g., [4, 5, 10–12]). How can we use the CMFs to obtain the cone fundamentals? This can be stated as a particular case of a more general transformation: the transformation of one set of primaries to another. Since CMFs $\bar{x}(\lambda)$, $\bar{y}(\lambda)$, and $\bar{z}(\lambda)$ are theoretically linear combinations of the cone fundamentals $\bar{l}(\lambda)$, $\bar{m}(\lambda)$, and $\bar{s}(\lambda)$, using matrix algebra we have:

$$\begin{pmatrix} \bar{x}(\lambda) \\ \bar{y}(\lambda) \\ \bar{z}(\lambda) \end{pmatrix} = \begin{pmatrix} K_p x_{pc} & K_d x_{dc} & K_t x_{tc} \\ K_p y_{pc} & K_d y_{dc} & K_t y_{tc} \\ K_p z_{pc} & K_d z_{dc} & K_t z_{tc} \end{pmatrix} \begin{pmatrix} \bar{l}(\lambda) \\ \bar{m}(\lambda) \\ \bar{s}(\lambda) \end{pmatrix} \tag{1}$$

where x_{pc}, y_{pc}, and z_{pc} represent the chromaticity coordinates of the protanope confusion point; x_{dc}, y_{dc}, and z_{dc} represent those of the deuteranope confusion point; and x_{tc}, y_{tc}, and z_{tc} represent those of the tritanope confusion point (K_p, K_d, and K_t are constant factors that scale the cone fundamentals). From this linear equation, we can get the cone fundamentals from the confusion points obtained from dichromatic observers:

$$\begin{pmatrix} \bar{l}(\lambda) \\ \bar{m}(\lambda) \\ \bar{s}(\lambda) \end{pmatrix} = \begin{pmatrix} K_p x_{pc} & K_d x_{dc} & K_t x_{tc} \\ K_p y_{pc} & K_d y_{dc} & K_t y_{tc} \\ K_p z_{pc} & K_d z_{dc} & K_t z_{tc} \end{pmatrix}^{-1} \begin{pmatrix} \bar{x}(\lambda) \\ \bar{y}(\lambda) \\ \bar{z}(\lambda) \end{pmatrix} \tag{2}$$

A more detailed description can be found in Section 8.2.5 of Ref. [4] and in Appendix, Part III of Ref. [5]. **Figure 2A** represents the set of cone fundamentals of normal trichromats proposed by Smith and Pokorny in 1975 [13] that was derived using this procedure. The spectral sensitivity of L, M, or S cones has been eliminated in **Figure 2B–D** to illustrate protanopia, deuteranopia, and tritanopia, respectively. More recently, Stockman and Sharpe [11, 14] have proposed a set of cone fundamentals based on the Stiles and Burch 10° CMFs (adjusted to 2°) and on direct measurements of the spectral sensitivity of dichromats, which has become the basis of the CIE proposal for physiologically relevant CMFs (http://www.cvrl.org/).

1.2. Color appearance: What are the colors seen by dichromats?

Each point represented in a chromaticity diagram represents metameric stimuli for normal trichromats, that is, color stimuli that presented at the appropriate intensity are perceptually

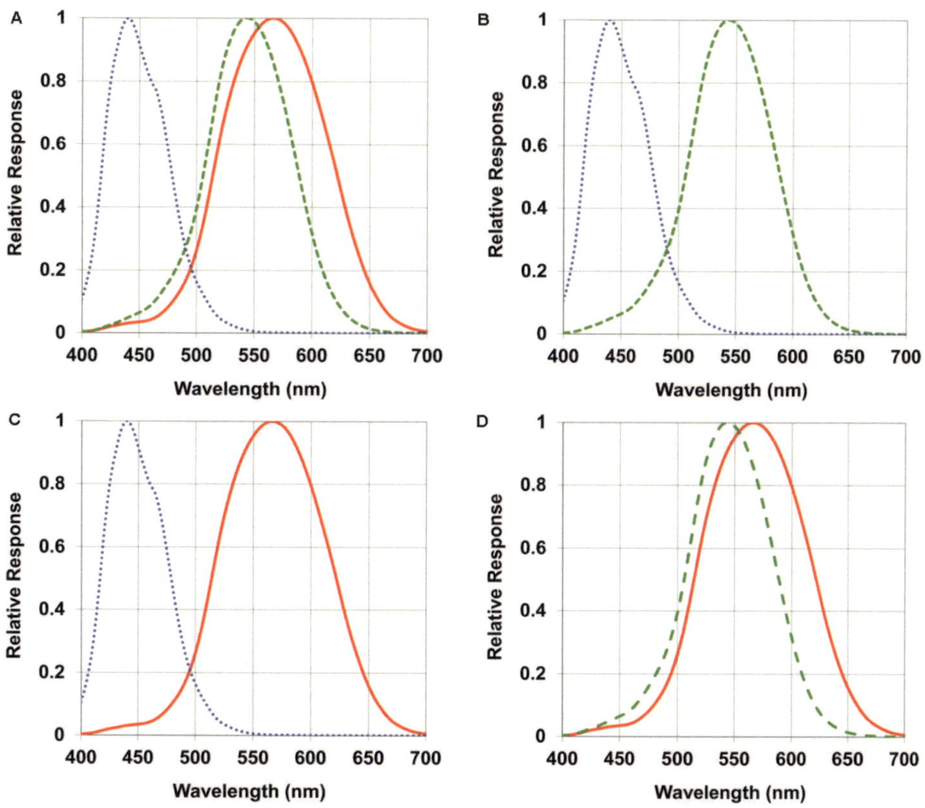

Figure 2. Cone fundamentals in normal trichromats and dichromats. **Figure 2A** represents the Smith and Pokorny cone fundamentals established in 1975 [13] normalized separately to peak at unity (these fundamentals were derived from the Judd CMFs published in 1951, a corrected version of the CIE 1931 CMFs that was later improved by Vos in 1978; see, e.g., Ref. [14]). L (solid red line), M (dashed green line), or S (dotted blue line) spectral sensitivity has been eliminated in **Figure 2B–D** to illustrate protanopia, deuteranopia, and tritanopia, respectively.

identical despite being physically different. As explained in Section 1.1.1, dichromats have more metamers than normal trichromats: for a given stimulus, the corresponding confusion line represents all the pseudoisochromatic stimuli that, if presented at the appropriate intensity, are metamers for a given type of dichromat. One of the most interesting and controversial aspects in color science has been the quest for an answer to this question: Which are the colors seen by dichromats? [2, 5]. More specifically, for a given color stimulus, which of all the possible metamers represents the actual color seen by a given type of dichromat?

In 1995, Viénot, Brettet, Ott, Ben M'Barek, and Mollon published a paper trying to respond to this question. Although the authors were aware that the quality of the sensations of other people cannot be fully known (page 128 in Ref. [15]), they described the rationale of an algorithm designed to simulate the chromatic appearance for a dichromat, in order to allow a normal trichromat to experience the same colors as a type of dichromat when looking at the same scene. The algorithm used to obtain this kind of simulations is described in more detail in Ref. [6]. The main assumption of the simulation is the aforementioned König hypothesis, that is, that dichromacy is a reduced version of normal trichromacy. On the basis of this assumption, the algorithm establishes, for a given color stimulus **Q**, and for a given type of dichromat, a stimulus **Q′** which is indistinguishable from **Q** and that produces the same color experience both in normal trichromats and dichromats of the specified type. In this way, it is possible to simulate the appearance of any scene, no matter how complex it is, for each type of dichromat, as it can be observed in **Figure 3**.

The algorithm proposed by Brettel et al. [6] does not use chromaticity diagrams to implement the simulation but LMS colorimetry (in which each tristimulus value represents the response of L, M, or S cones). Specifically, the authors used the cone fundamentals proposed by Stockman, MacLeod, and Johnson in 1993 [16]. In this physiologically based cone color space, **Q** and **Q′** stimuli are defined as **Q** = (L_Q, M_Q, S_Q) and **Q′** = ($L_{Q'}$, $M_{Q'}$, $S_{Q'}$).

Figure 3. Dichromatic color appearance simulations. (A) Original Joaquin Sorolla's painting *Saliendo del baño* ("Coming out of the bath"); (B) protanopia simulation; (C) deuteranopia simulation; and (D) tritanopia simulation. The simulations were performed using the color simulation tool named "Vischeck" (http://www.vischeck.com/). Note that the great changes in hue from the pinkish colors in the original image (A) toward yellowish colors in the protanopia (B) and deuteranopia (C) simulation do not occur in the tritanopia simulation (D) (according to the model of Brettel et al. [6], reddish and greenish blue are identical for tritanopes and normal trichromats). Because of possible errors in reproduction, the actual colors could present some differences with the original images.

In a three-dimensional representation of this space, each axis represents the response of one of the three types of cones, and confusion lines are easily determined as parallel lines to one of the axes, since they represent all the stimuli that only vary in the response of the cone that is absent for a given type of dichromat, keeping constant the response of the two remaining cones: confusion lines for protanopes are parallel to L axis ($M_{Q'} = M_Q$ and $S_{Q'} = S_Q$), confusion lines for deuteranopes are parallel to M axis ($L_{Q'} = L_Q$ and $S_{Q'} = S_Q$), and confusion lines for tritanopes are parallel to S axis ($L_{Q'} = L_Q$ and $M_{Q'} = S_Q$).

But only one of the stimuli contained in a given confusion line, the one defined by the vector **Q′**, would fulfill the requirement of producing the *same* experience (according to the authors) in dichromats and normal trichromats. Which one is the correct stimulus for each type of dichromat ($L_{Q'}$, $M_{Q'}$, and $S_{Q'}$ in protanopes, deuteranopes, and tritanopes respectively)? Given this uncertainty, Brettel et al. [6] drew on data collected from unilateral dichromats (observers who are dichromats in one eye and supposedly normal trichromats in the other) in order to establish the stimuli that cause common chromatic experiences both for dichromats and normal trichromats. Which are the stimuli that serve as perceptive anchors to translate the colors of any image to the colors supposedly seen by dichromats? In brief:

1. It is assumed that equal-energy stimulus **aE** ($0 \leq a \leq 1$) produces the same perceptual response both in dichromats and normal trichromats, regardless of its intensity (the authors defined **E** as the metamer of the equal-energy stimulus with the maximum reproducible intensity in the screen where the simulation was run). Therefore, the origin **O** ($L_O = M_O = S_O = 0$) and the equal-energy stimulus **E** define the neutral axis *OE*, perceived as achromatic.

2. The three-dimensional LMS space of normal trichromats is reduced to two hemiplanes in dichromats, each one of them associated with the perception of one hue. Which are the hues equally perceived by dichromats and normal trichromats? From the published studies on unilateral dichromats, it was inferred that:

   2.1. The hues produced by spectral stimuli of 575 nm (yellow) and 475 nm (blue) are identical for protanopes, deuteranopes, and normal trichromats. This inference was based on several studies on unilateral dichromats of genetic origin.

   2.2. The hues produced by spectral stimuli of 660 nm (red) and 485 nm (greenish blue) are identical for tritanopes and normal trichromats. This inference was based on one study on a case of acquired unilateral tritanopia.

The hemiplanes that contain the LMS coordinates related to the chromatic experiences common to dichromats and normal trichromats are defined by the achromatic axis *OE* and by the LMS coordinates of 575 and 475 nm in the case of protanopes and deuteranopes or by the achromatic axis *OE* and by the LMS coordinates of 660 and 485 nm in the case of tritanopes. Stimulus **Q′** is defined by the intersection of the confusion line of **Q** and the corresponding hemiplane. This procedure can be applied to all the elements of any image in order to translate its original colors to the colors supposedly seen by dichromats according to the aforementioned assumptions of the model.

Despite the reservations expressed by the authors concerning the validity of the method, they defended its utility for normal trichromats to evaluate the chromatic variety experienced by dichromats when looking at any scene.

1.3. Dataset

After the previous theoretical and technical background, the following sections will present some empirical data relating colorimetric variables with dichromatic experiences. In Section 2, we will present an experimental method developed to evaluate the accuracy of color simulation tools in relation to dichromats [17]. Section 3 gives an overview of two competitive models developed to describe and predict the confusions made by red-green dichromats when naming colors [18, 19]. Both models are based on colorimetric variables. Model A is based on the activity of the yellow-blue and the achromatic mechanisms. Model B includes a third variable: the aforementioned residual red-green discrimination. Section 4 deals with an emotional aspect related to color, describing color preferences in red-green dichromats and comparing them to the preferences of normal trichromats [20]. Finally in Section 5, we will summarize the most important conclusions of our research on red-green dichromats.

2. Evaluation of color appearance simulation tools

2.1. Color appearance simulation tools

Several color appearance simulation tools are available. All of them take an original image as input and apply some image processing algorithm or optical filtering to get a processed or filtered image that represents the colors supposedly seen by observers with different types of color vision deficiencies. For example, dichromatic color appearance simulation tools translate any image into versions of that image to simulate protanopia, deuteranopia, and tritanopia, as illustrated in **Figure 3**. Simulation tools also can be used to create systems for avoiding color combinations difficult to discriminate by color anomalous observers [21]. Our research on color appearance simulation tools, to be described in Section 2.2, has been focused on color appearance simulation tools for the most common types of dichromatic color vision: protanopia and deuteranopia.

2.2. Simulcheck: A method to evaluate the accuracy of dichromatic color appearance simulation tools

Figure 3 shows four versions of the famous masterpiece *Saliendo del baño* ("Coming out of the bath") painted by the Spanish painter Joaquin Sorolla in 1915. The one at the upper left (**Figure 3A**) is the original image. **Figure 3B–D** is three versions of that image created to mimic how the original picture is seen by the three types of dichromats, protanopes, deuteranopes, and tritanopes, respectively. The simulation was performed by means of the Internet available color simulation tool named "Vischeck" (which is based on the algorithm of Brettel et al. [6] described in Section 1.2) (http://www.vischeck.com/). **Figure 3C** is reproduced again in **Figure 4A** to facilitate its comparison with the simulation of deuteranopia provided by "Coblis" (**Figure 4B**), another color simulation tool (http://www.color-blindness.com/coblis-color-blindness-simulator/). It is easy to appreciate some important differences between both deuteranopia simulations. For example, the **Figure 3A** whites (i.e., the woman's shirt, the towel or the waves' crest) become light pinks in **Figure 4B** but not in **Figure 4A**. It means that the same input (**Figure 3A**) produced different colors (**Figure 4A** and **B**) when using different tools to mimic the same type of dichromatic color vision (deuteranopia). Obviously, it can be concluded that at least one of the tools is not performing accurately.

Lillo et al. created and developed in 2014 [17] a method named "Simulcheck" to evaluate ("-check") the accuracy of color simulation ("Simul-") tools. This method is based on the measurement of two colorimetric variables, h_{uv} (hue angle) and L_R (relative luminance). Each one of these variables is related to a relevant feature of dichromatic color vision, respectively (see below): the existence of pseudoachromatic colors and the differences in the lightness perception for those colors.

h_{uv}, a standard colorimetric variable specified in the *CIE 1976 u′v′* chromaticity diagram that correlates with hue [1], was used to measure the accuracy of color simulation tools in relation to pseudoachromatic colors. The hue angle of a stimulus is simply calculated as the angle formed by an imaginary horizontal line passing through the reference white and the line defined by the reference white and the stimulus chromaticity coordinates. As it was previously commented, **Figure 1** right panel shows several *CIE 1976 u′v′* chromaticity diagrams including some confusion lines that represent pseudoisochromatic stimuli. One of these lines (the solid line) passes through the chromaticity of the equal-energy stimulus (the point labeled with the letter E). It means that for the specific type of dichromacy considered (protanopia in the upper diagram, deuteranopia in the central diagram, and tritanopia in the bottom diagram) every stimuli represented by a point in this line must be perceived as an achromatic color (white, gray, or black) like the one seen when looking at an equal-energy stimulus. Because these stimuli are achromatic for a type of dichromat but not for normal trichromats, they are frequently named "pseudoachromatics" [7].

Figure 4. Two color appearance simulations for deuteranopia of a Joaquin Sorolla's painting. (A) Vischeck simulation and (B) Coblis simulation.

Let us use the right panel of the first row in **Figure 1** (representing confusion lines for protanopes in the *CIE 1976 u′v′* chromaticity diagram) to make an intuitive explanation on how the hue angle (h_{uv}) is used by Simulcheck to measure the accuracy of color simulation tools in relation to pseudoachromatic colors. The pseudoachromatic line for protanopes intersects with the chromaticity diagram (1) between 490 and 495 nm (the so-called neutral point for protanopes, represented by a black triangle, see, e.g., Ref. [12]) and (2) in the purple line near 700 nm. The corresponding h_{uv} values are (1) near 184° that corresponds to stimuli usually perceived as green (bluish green, actually) by normal trichromats, so these stimuli can be named "protanope pseudoachromatic greens," and (2) near 4° that corresponds to stimuli usually perceived as red or pink by normal trichromats, so they can be named "protanope pseudoachromatic reds."

Regarding the differences in lightness perception for pseudoachromatics, such differences derive from fact that the spectral sensitivity of protanopes and deuteranopes differs from that of normal trichromats (see, e.g., Ref. [2]). As a consequence, their lightness perception is also different (see next section). For protanopes, reds are darker and greens are lighter than for normal trichromats. For deuteranopes the pattern is exactly the reverse.

2.2.1. Methods

Participants. Ten normal trichromats and 10 dichromats (5 protanopes, 5 deuteranopes) were participated. Their color vision was assessed with a set of pseudoisochromatic tests and a Nagel anomaloscope (Tomey AF-1, Tomey, Nagoya, Japan). All were native Spanish speakers.

Stimuli. Two chromatic stimulus sets and an achromatic stimulus set were used. Each chromatic set included 40 stimuli evenly spaced in h_{uv}. The first set included stimuli with varying chroma and lightness (maximum chroma set), whereas the second set included stimuli with equal chroma and lightness (constant chroma set). The achromatic set included 20 achromatic stimuli evenly spaced in lightness from black to white.

Procedure. Simulcheck includes two tasks: the "pseudoachromatic stimuli identification task" and the "minimum achromatic contrast task." The first task measures the h_{uv} values of the pseudoachromatic stimuli selected by the observers from a set of colors spanning the whole hue circle (see above): participants are required to select the stimulus more similar to gray. Simulcheck includes a "minimum achromatic contrast task" to measure L_R (relative luminance), a psychophysical variable directly related to lightness perception. This task estimates L_R (relative luminance), a psychophysical variable directly related to lightness perception. The pseudoachromatic stimulus selected in the first task is presented as text over the 20 stimuli of the achromatic set. Participants are required to select the stimulus which makes it more difficult to read the text which is the one with minimum contrast with the pseudoachromatic stimuli. L_R is computed as the ratio between L_T (the luminance of the achromatic background selected) and L (the standard luminance values of the pseudoachromatic stimuli).

2.2.2. Results and conclusions

We compared the results obtained by real red-green dichromats and simulated dichromats (i.e., normal trichromats performing the tasks with the transformed stimuli). Mann-Whitney

U tests were performed and *p* values lower than .05 indicated significant differences between real and simulated dichromats (see Figures 9 and 10 in Ref. [17]). Hue angle values were very similar for real red-green dichromats (protanopes and deuteranopes) and for those simulated by Vischeck, but not for those simulated by Coblis. As it happened for h_{uv}, the L_R values of the colors simulated by Vischeck were very much more accurate than their Coblis' equivalents. Therefore we concluded [17] that Vischeck is a much more accurate tool than Coblis to simulate red-green dichromacy.

3. Color naming models in dichromats

Much of the modern work on color and language [22–24] has been inspired by the seminal work of Berlin and Kay [25] about the basic color terms (BCTs). These terms are monolexemic, known and used by all members of the language community and can be used to communicate about the color of any type of object. The general limitations in color naming of red-green dichromats [26] also appear for using BCTs [18].

Two models based on transformed colorimetric variables were developed by Moreira et al. in 2014 [19] to predict the confusions made by red-green dichromats when naming colors using basic color terms (BCTs). Castilian Spanish BCTs are 11 (English equivalents after hyphen): rojo-red, verde-green, amarillo-yellow, azul-blue, blanco-white, negro-black, naranja-orange, morado-purple, marrón-brown, rosa-pink, and gris-gray. Model A is based on the activity of the yellow-blue and the achromatic mechanisms. Model B includes a third variable: residual red-green discrimination.

In essence, Model A departs from the *reduction hypothesis*; therefore, it shares the main assumption of the model of Brettel et al. [6] described previously (see Section 1.2). More specifically, Model A is based on the same assumptions of the model of Brettel et al. [6] regarding the chromatic experiences shared by dichromats and normal trichromats, but it uses standard colorimetric variables instead of LMS colorimetry to describe and predict the use of BCTs (see below). In contrast, Model B includes residual *red-green discrimination*, given the great amount of research that evidences that the major part of red-green dichromats behaves as anomalous trichromats for large-field stimuli (over 3°) both in psychophysical tasks [8, 9] and color-naming tasks [27–32]. In fact, as it was stated before that the existence of residual red-green discrimination can challenge the use of confusion lines to predict the confusions made by red-green dichromats and casts serious doubts on the assumptions on which Model A is based (a critical reexamination can be seen in Ref. [33]).

A full description of Models A and B can be found in Ref. [19]. Here we will only present the essential features of both models. Model A is entirely based on the activity of the yellow-blue and the achromatic mechanisms, specified in terms of variables s' and L^*_T, respectively. Both of these variables are related to standard colorimetric variables derived from CIELUV space and its chromaticity diagram (*CIE 1976 u'v'*). Like standard variables, which do not consider individual differences within normal trichromats, Model A variables do not take into account individual differences within protanopes or deuteranopes.

The first variable of Model A, s', identifies hue and quantifies saturation for protanopes (s'_p) and deuteranopes (s'_d) using the *1976 CIE u'v'* chromaticity diagram. Specifically, s' is computed from the projection of the 2D chromaticity diagram onto a 1D representation of chromaticity related to the chromatic experiences shared by dichromats and normal trichromats according to the model of Brettel et al. [6]. **Figure 5** shows how s'_p identifies hue (yellow or blue) and quantifies saturation for protanopes using the *CIE 1976 u'v'* chromaticity diagram as follows (the same rationale applies to deuteranopes):

1. There is a biunivocal relation between confusion lines and variable s'_p. For example, $s'_p = 0$ only for the pseudoachromatic confusion line for protanopes, i.e., the protanopic confusion line passing through the chromaticity of the reference white (thick solid line in **Figure 5**; see Section 2.2). This confusion line includes all the stimuli that protanopes perceive as achromatic according to Model A. Any other confusion line intersects either with the "blue line" ($\lambda_D = 475$ nm) or with the "yellow line" ($\lambda_D = 575$ nm) at a given intersection point.

2. Variable s' is simply computed as the distance between the achromatic point and the intersection point (except for a scalar, this distance coincides with the value of the standard colorimetric variable, s_{uv} [1] for the intersection point). We arbitrarily decided s' to be positive for intersection points located on the "yellow line" and negative for those located on the "blue line."

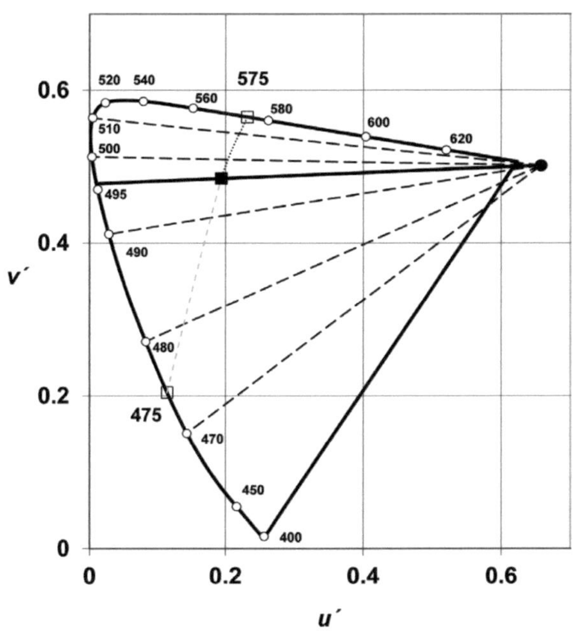

Figure 5. Intersection between confusion lines for protanopes and the "yellow" (black dotted line) or "blue" (grey dashed line) lines according to Model A. Every confusion line (dashed line) radiates from the protanopic confusion point (black circle, $u' = 0.658$, $v' = 0.501$) and intersects with either the yellow or the blue line (white squares show the chromatic coordinates of 475 and 575 nm). The solid line crossing the achromatic point (black square, sample S 0500-N of NCS atlas in our study, $u' = 0.19$, $v' = 0.48$; see Appendix A in Ref. [18]) includes pseudoachromatic stimuli (achromatic for protanopes). The s' variable was defined as the distance from the achromatic point to the intersection point between the confusion line of a given stimulus and either the yellow or the blue line (see text for details).

The second variable of Model A, transformed lightness, L^*_T, is computed as standard lightness (L^* [1]), but instead of using the standard luminance factor, it takes into account that protanopes lack L cones and deuteranopes lack M cones. Transformed protanope and deuteranope lightness, L^*_p and L^*_d, were computed as follows:

1. Transformed luminance factor for protanopes (Y_p) and deuteranopes (Y_d) was computed on the assumption that the luminance factor of the reference white is exactly the same for normal trichromats and dichromats, which makes it possible to compute a scaling factor for protanopes and other for deuteranopes. We used Stockman and Sharpe $L_2(\lambda)$ and $M_2(\lambda)$ cone spectral sensitivities [14] and derived the relative contribution of L and M cones to luminosity from $V_2^*(\lambda) = 1.5\, L_2(\lambda) + M_2(\lambda)$ (Eq. 7 in Ref. [14]).
2. Finally, the corresponding transformed lightness values for protanopes and deuteranopes (L^*_p and L^*_d) were computed.

Model B includes s' and L^*_T variables as Model A but also includes a third variable, ΔRG_{res}, to take residual activity in the R-G channel into account. According both to Model A and Model B, the probability of using a given BCT decreases insofar as a stimulus differs from the best exemplars of the corresponding category in terms of s' and L^*_T. But only according to Model B, this probability also decreases as the stimulus moves away from the focal color of that category in terms of incremental distance along the corresponding confusion line, ΔRG_{res}.

3.1. Methods

Participants. 32 normal trichromats (17 females, 15 males) and 17 dichromats (8 protanopes, 9 deuteranopes) were participated. All were native Spanish speakers, and they were assessed with a set of color vision tests and a Raleigh match on a Nagel anomaloscope (Tomey AF-1, Tomey, Nagoya, Japan).

Stimuli. 102 stimuli from the NCS color atlas were used. They included (1) the best exemplar of each basic color category in Spanish (a detailed description on the equivalent basic color terms can be found in Section 3), (2) stimuli in the boundary of neighboring basic color categories, and (3) stimuli halfway between a boundary and the best exemplar of a basic color category.

Procedure. Stimuli were presented simultaneously, and participants were asked to select instances (mapping task) or to select the prototype (best exemplar task) of a given basic color term. Both tasks were performed independently.

3.2. Results and conclusions

The comparison of the predictions of Models A and B and the observed responses of protanopes and deuteranopes showed that Model B was more accurate than Model A in predicting the use of BCTs, indicating that *residual red-green discrimination* must be taken into account in order to explain the use of BCTs by red-green dichromats, besides the yellow-blue and the achromatic mechanisms. For example, for the use of green by protanopes, the proportion of variance (R^2) accounted for by Model A was 56%, whereas the proportion of variance accounted for by Model B was 96% (i.e., the increment in the proportion of variance accounted for by using Model B instead of Model A was $\Delta R^2 = 40\%$) (see Table VI in Ref. [19] for a systematic comparison of both models).

4. Color preference in dichromats

Color preference in normal observers [34] has been explained considering object-color associations [35], emotional factors [36], and the responses in the chromatic opponent mechanisms [37]. For the last case, standard colorimetric measurements have been used to estimate the response magnitudes in the three retinal cones (L, M, and S) and the two opponent mechanisms (red-green, L-M, and yellow-blue, S-(L + M)). Álvaro et al. [20] studied color preferences in dichromats and compared them to the preferences of normal trichromats (**Figure 6**; see also Schloss [38]).

4.1. Methods

Participants. 32 normal trichromats (17 females, 15 males) and 32 dichromats (15 protanopes, 17 deuteranopes) were participated. All were assessed with a set of color vision tests and a Nagel anomaloscope (Tomey AF-1, Tomey, Nagoya, Japan), and they were native Spanish speakers.

Stimuli. Three stimulus sets were saturated, light and dark versions of red, orange, yellow, chartreuse, green, cyan, blue, and purple. Those stimuli were approximations of the Berkley Color Project set [39].

Procedure. Participants were required to rate their preference (color-preference task) or to name aloud (color-naming task) the color individually presented. Preference rating scale

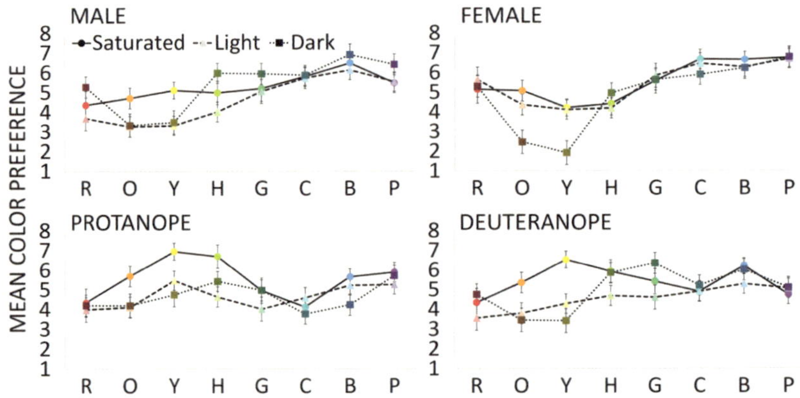

Figure 6. Mean preference ratings averaged for trichromatic males and females and dichromatic protanopes and deuteranopes. Mean preference ratings (± SEM) are shown as a function of hue, given by the x-axis: red (R), orange (O), yellow (Y), chartreuse (H), green (G), cyan (C), blue (B), and purple (P). Separate lines represent colors from the saturated (solid with circles), light (dashed with triangles), and dark (dotted with squares) sets. Marker colors are only an approximation of those in the experiment. This figure is a reproduction of **Figure 1** from Álvaro et al. [20].

ranged from 1 to 10. Color names were restricted to the 11 Spanish basic color terms (see Section 3). Both tasks were performed independently.

4.2. Results and conclusions

We compared the color preference ratings of dichromats and normal trichromats (**Figure 6**) by means of a mixed-model ANOVA with set (saturated, light, and dark) and color (red, orange, yellow, chartreuse, green, cyan, blue, and purple) as within-subject factors and group (trichromat males, trichromat females, protanopes, and deuteranopes) as the between-subjects factor (*p* values lower than .05 indicated significant differences). While normal trichromats' preference peaks at blue and hits bottom at yellow-green [35, 37, 39], red-green dichromats' preference peaks at yellow and the preference for blue is weaker than normal trichromats'. Protanope's preference was more affected than deuteranope's preference. In relation to the underlying mechanisms to color preference, we entered cone contrasts (between color stimuli and background; see **Figure 7**) as predictors in linear regressions analyses with average color preference ratings (**Figure 6**) as the dependent variable. Cone contrast partially explained normal trichromat's preference with the yellow-blue system being the most important. Variations of cone-contrast theory partially explained protanope's preference with again a special relevance of the yellow-blue system activation (bottom of **Figure 7**) but showing evidence of red-green residual activity in deuteranopes

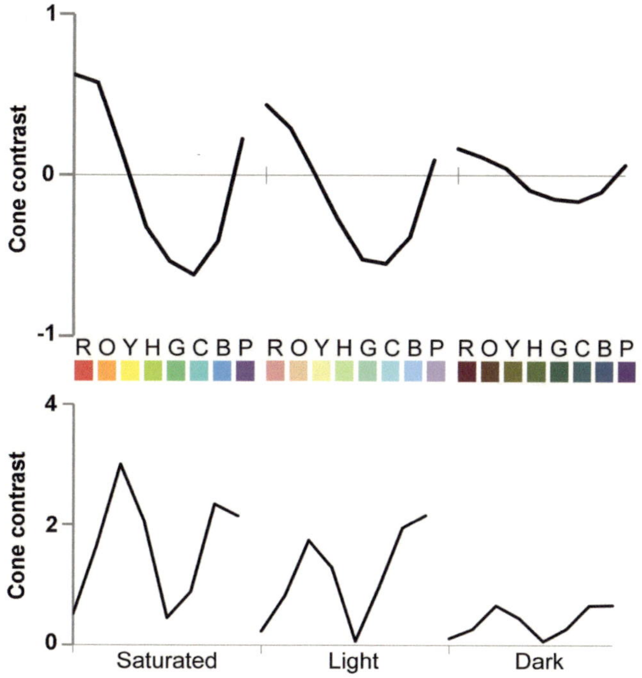

Figure 7. Cone contrast of the stimulus sets used to evaluate color preference in normal trichromats and red-green dichromats. Red-green (L − M, top panel) and yellow-blue activation (|S−(L + M)|, bottom panel) stimulus-background cone-contrast values for the saturated, light, and dark set of colors used in Ref. [20]: red (R), orange (O), yellow (Y), chartreuse (H), green (G), cyan (C), blue (B), and purple (P). This figure has been adapted from **Figure 3** in Álvaro et al. [20].

(top of **Figure 7**). For example, the activation of the yellow-blue system was a significant predictor and explained almost two thirds of the variance (R^2 = .61) in protanopes' preference for the full color set and more than three quarters of the preference for saturated (R^2 = .76) and light (R^2 = .88) colors (see Table 1 in Ref. [20]). Besides, this novel study demonstrated that the fluency of processing also underlies males' preference (either normal trichromats or dichromats): there was a higher preference for the colors named more accurately, quicker, more consistently and with greater consensus in normal trichromatic males and dichromatic males but not in normal trichromatic females.

5. Conclusions

The main conclusions of our research on red-green dichromacy related to colorimetry can be stated as follows:

1. Simulcheck is a valid method to evaluate color simulation tools in relation to dichromacy [17]. Hue angle and relative luminance are the two colorimetric dependent variables measured by the two psychophysical tasks included in the method. The similarity of the pattern of results obtained by simulated and real dichromats is a good indicator of the validity and the accuracy of color simulation tools.
2. Residual red-green discrimination must be taken into account to explain the use of basic color terms (BCTs) by red-green dichromats, as well as yellow-blue and achromatic mechanisms [19]. For most part of dichromats, distance along confusion lines (besides yellow-blue and achromatic mechanisms) is a relevant colorimetric variable for the use of BCTs.
3. Color preference of red-green dichromats is different from color preference of normal trichromats [20]. Yellow-blue cone contrast (computed from colorimetric variables) accounts for dichromats' pattern of preference, with some evidence for residual red-green activity in deuteranopes' preference.

Acknowledgements

This work was partly funded by a Ministerio de Economía y Competitividad Grant PSI2017-82520.

Author details

Humberto Moreira[1]*, Leticia Álvaro[2], Anna Melnikova[1] and Julio Lillo[1]

*Address all correspondence to: humbermv@psi.ucm.es

1 Complutense University, Madrid, Spain

2 University of Sussex, Falmer, United Kingdom

References

[1] Hunt RWG, Pointer R. Measuring Colour. 4th ed. Chichester: John Wiley & Sons; 2011. DOI: 10.1002/9781119975595

[2] Sharpe LT, Stockman A, Jägle H, Nathans J. Opsin genes, cone photopigments, color vision, and color blindness. In: Gegenfurtner KR, Sharpe LT, editors. Color Vision: From Genes to Perception. Cambridge: Cambridge University Press; 1999. p. 3-51

[3] Neitz J, Neitz M. The genetics of normal and defective color vision. Vision Research. 2011;**51**(7):633-651. DOI: 10.1016/j.visres.2010.12.002

[4] Wyszecki G, Stiles WS. Color Science: Concepts and Methods, Quantitative Data and Formulae. 2nd ed. New York: John Wiley & Sons; 1982

[5] Kaiser PK, Boynton RM. Human Color Vision. Washington, DC: Optical Society of America; 1996

[6] Brettel H, Viénot F, Mollon JD. Computerized simulation of color appearance for dichromats. Journal of the Optical Society of America A. 1997;**14**(10):2647-2655. DOI: 10.1364/JOSAA.14.002647

[7] Birch J. Diagnosis of Defective Colour Vision. 2nd ed. Oxford: Butterworth-Heinemann; 2001

[8] Smith VC, Pokorny J. Large-field trichromacy in protanopes and deuteranopes. Journal of the Optical Society of America. 1977;**67**(2):213-220. DOI: 10.1364/JOSA.67.000213

[9] Nagy AL. Large-field substitution Rayleigh matches of dichromats. Journal of the Optical Society of America. 1980;**70**:778-784. DOI: 10.1364/JOSA.70.000778

[10] Mollon JD. The origins of modern color science. In: Shevell SK, editor. The Science of Color. 2nd ed. Oxford: Optical Society of America; 2003. p. 1-39

[11] Stockman A, Sharpe LT. Cone spectral sensitivities and color matching. In: Gegenfurtner KR, Sharpe LT, editors. Color Vision: From Genes to Perception. Cambridge: Cambridge University Press; 1999. p. 53-87

[12] Smith VC, Pokorny J. Color matching and color discrimination. In: Shevell SK, editor. The Science of Color. 2nd ed. Oxford: Optical Society of America; 2003. p. 103-148

[13] Smith VC, Pokorny J. Spectral sensitivity of the foveal cone photopigments between 400 and 500 nm. Vision Research. 1975;**15**(2):161-171. DOI: 10.1016/0042-6989(75)90203-5

[14] Stockman A, Sharpe LT. The spectral sensitivities of the middle- and long-wavelength-sensitive cones derived from measurements in observers of known genotype. Vision Research. 2000;**40**(13):1711-1737. DOI: 10.1016/S0042-6989(00)00021-3

[15] Viénot F, Brettel H, Ott L, Ben M'Barek A, Mollon JD. What do color-blind people see. Nature. 1995;**376**(6536):127-128. DOI: 10.1038/376127a0

[16] Stockman A, MacLeod DIA, Johnson NE. Spectral sensitivities of the human cones. Journal of the Optical Society of America A. 1993;**10**(12):2491-2521. DOI: 10.1364/JOSAA.10.002491

[17] Lillo J, Álvaro L, Moreira H. An experimental method for the assessment of color simulation tools. Journal of Vision. 2014;**14**(8):1-19. DOI: 10.1167/14.8.15

[18] Lillo J, Moreira H, Álvaro L, Davies I. Use of basic color terms by red-green dichromats: 1. General description. Color Research & Application. 2014;**39**(4):360-371. DOI: 10.1002/col.21803

[19] Moreira H, Lillo J, Álvaro L, Davies I. Use of basic color terms by red-green dichromats. II. Models. Color Research & Application. 2014;**39**(4):372-386. DOI: 10.1002/col.21802

[20] Álvaro L, Moreira H, Lillo J, Franklin A. Color preference in red–green dichromats. Proceedings of the National Academy of Sciences of the United States of America. 2015;**112**(30):9316-9321. DOI: 10.1073/pnas.1502104112

[21] Hassan MF, Paramesran R. Naturalness preserving image recoloring method for people with red–green deficiency. Signal Processing: Image Communication. 2017;**57**(Supplement C):126-133. DOI: https://doi.org/10.1016/j.image.2017.05.011

[22] Lindsey DT, Brown AM. The color lexicon of American English. Journal of Vision. 2014; **14**(2):17-17. DOI: 10.1167/14.2.17

[23] Gibson E, Futrell R, Jara-Ettinger J, Mahowald K, Bergen L, Ratnasingam S, et al. Color naming across languages reflects color use. Proceedings of the National Academy of Sciences of the United States of America. 2017;**114**(40):10785-10790. DOI: 10.1073/pnas.1619666114

[24] Kuriki I, Lange R, Muto Y, Brown AM, Fukuda K, Tokunaga R, et al. The modern Japanese color lexicon. Journal of Vision. 2017;**17**(3):1-1. DOI: 10.1167/17.3.1

[25] Berlin B, Kay P. Basic Color Terms: Their Universality and Evolution. Berkeley: University of California Press; 1969

[26] Nagy BV, Németh Z, Samu K, Ábrahám G. Variability and systematic differences in normal, protan, and deutan color naming. Frontiers in Psychology. 2014;**5**(1416). DOI: 10.3389/fpsyg.2014.01416

[27] Boynton RM, Scheibner HM. On the perception of red by red-blind observers. Acta Chromatica. 1967;**1**:205-220

[28] Scheibner HM, Boynton RM. Residual red-green discrimination in dichromats. Journal of the Optical Society of America. 1968;**58**(8):1151-1158. DOI: 10.1364/JOSA.58.001151

[29] Nagy AL, Boynton RM. Large-field color naming of dichromats with rods bleached. Journal of the Optical Society of America. 1979;**69**:1259-1265. DOI: 10.1364/JOSA.69.001259

[30] Montag ED, Boynton RM. Rod influence in dichromatic surface color perception. Vision Research. 1987;**27**(12):2153-2162. DOI: 10.1016/0042-6989(87)90129-5

[31] Montag ED. Surface color naming in dichromats. Vision Research. 1994;**34**(16):2137-2151. DOI: 10.1016/0042-6989(94)90323-9

[32] Uchikawa K. Trichromat-like categorical color naming of dichromats. Vision: The Journal of the Vision Society of Japan. 2008;**20**:62-66

[33] Broackes J. Unilateral colour vision defects and the dimensions of dichromat experience. Ophthalmic and Physiological Optics. 2010;**30**(5):672-684. DOI: 10.1111/j.1475-1313.2010.00774.x

[34] Schloss KB, Palmer SE. An ecological framework for temporal and individual differences in color preferences. Vision Research. 2017. DOI: https://doi.org/10.1016/j.visres.2017.01.010

[35] Taylor C, Franklin A. The relationship between color-object associations and color preference: Further investigation of ecological valence theory. Psychonomic Bulletin & Review. 2012;**19**:190-197. DOI: 10.3758/s13423-012-0222-1

[36] Sato K, Inoue T. Perception of color emotions for single colors in red-green defective observers. PeerJ. 2016;**4**:e2751. DOI: 10.7717/peerj.2751

[37] Hurlbert AC, Ling Y. Biological components of sex differences in color preference. Current Biology. 2007;**17**:R623-R625. DOI: 10.1016/j.cub.2007.06.022

[38] Schloss KB. Color preferences differ with variations in color perception. Trends in Cognitive Sciences. 2015;**19**(10):554-555. DOI: 10.1016/j.tics.2015.08.009

[39] Palmer SE, Schloss KB. An ecological valence theory of human color preference. Proceedings of the National Academy of Sciences of the United States of America. 2010;**107**: 8877-8882. DOI: 10.1073/pnas.0906172107

Chapter 2

Image Segmentation Based on Mathematical Morphological Operator

Jianjun Chen, Haijian Shao and Chunlong Hu

Additional information is available at the end of the chapter

http://dx.doi.org/10.5772/intechopen.72603

Abstract

Image segmentation is the process of partitioning a digital image into multiple regions (sets of pixels); the pixels in each region have similar attributes. It is often used to separate an image into regions in terms of surfaces, objects, and scenes, especially for object location and boundary extraction. Until now, many general-purpose algorithms and techniques have been proposed for image segmentation. Typical and traditional methods are: (1) threshold-based method; (2) edge-based method; and (3) region-based method. In this chapter, we propose an approach of image segmentation based on mathematical morphology operator: toggle operator. The experimental result shows that the proposed method can segment natural scene images into homogeneous regions effectively.

Keywords: homogeneous region, image segmentation, mathematical morphology, toggle operator

1. Introduction

Image segmentation is typically used to partition an image into meaningful parts. Thus, it has a significant application in image analysis and understanding. The result of image segmentation is a set of regions (each region is a set of pixels) that collectively cover the entire image, or a set of contours (i.e., boundaries, consisting of lines, curves, etc.) extracted from the image. The pixels in one region have similar characteristics in terms of color, intensity, or texture. Adjacent regions are significantly different with respect to the same characteristic(s) [1, 2].

Until now, a great variety of algorithms have been proposed for image segmentation [3, 4]. These methods are generally classified into three categories: thresholding segmentation, edge-based segmentation, and region-based segmentation. Each method has its own advantages and disadvantages; there is no single method that can be applied effectively for segmenting all kinds of images. This is because different images have different features and properties.

Therefore, for different images, the segmentation requires different techniques. Images can be divided into the following categories according to their attributes and characteristics. From the color point of view, images include grayscale images, binary images, and color images; from the texture point of view, images include texture images and nontexture images. Based on the image features, the following subsections will consider some proposed classical approaches in more detail.

1.1. Threshold-based segmentation

Thresholding segmentation is a pixel-based method for image segmentation. It is the simplest method based on the variation of intensity between the object pixels and background pixels. Therefore, it is often used to separate out regions of an image corresponding to objects that we are interested in.

In order to differentiate the pixels that are located in the region of interest from the rest, a comparison is performed for each pixel intensity value with respect to a threshold. In this method, pixels are divided into two classes that are typically named "foreground" and "background." Pixels with values less than threshold are placed in one class, and the rest are placed in the other class. Therefore, this method is often used to convert a grayscale image into a binary image.

$$I_{bw}(x,y) = \begin{cases} 1 & I_{gray}(x,y) \geq T \\ 0 & I_{gray}(x,y) < T \end{cases} \tag{1}$$

where I_{gray} is the grayscale image, I_{bw} is the binary image, (x, y) is the coordinate of target pixel, and T is the threshold value. This method is most effective for images with high levels of contrast. However, the key of this method is to select the well-suited threshold value.

Many researchers contribute to the work for automated selecting the threshold T based on the computer. Their thresholding methods can be categorized into the following groups based on the information the algorithm manipulates [5–9]:

(1) Histogram shape-based methods: the peaks, valleys, and curvatures of the smoothed histogram are analyzed.

  a. Double-peak threshold method: suppose the histogram of the image is a bimodal distribution (regions with uniform intensity give rise to strong peaks in the histogram), then the value of valley point can be chosen as the threshold.

  b. Minimum variance method: suppose that a region has relatively homogeneous gray values, it will make sense to select a threshold that minimizes the variance of the gray values within regions.

  c. Maximum variance method (usually called the Otsu method [10]): chooses a good threshold value by maximizing the variance between objects and background. The histogram can be divided into two classes, while the interclasses' variance is maximized.

(2) Clustering method: representing the image as a set of clusters, an ideal threshold value is determined by iteratively reducing the clusters until there are two clusters left. The two remaining clusters are the background and the foreground (object).

(3) Maximum-entropy-based method: determines an ideal threshold value by maximizing entropy for the probability distribution of the foreground and background regions, which are represented in the histogram.

(4) Local thresholding method: chooses the threshold value for each pixel according to the local image characteristics. In this method, a different threshold value is selected for each pixel in the image.

(5) Watershed method: considers the gradient magnitude of an image as a topographic surface where high gradient denotes peaks, while low gradient denotes valleys. Start by filling every isolated valley with different colored water. As the water rises, water from different valleys will start to merge. To avoid that, barriers are built in the locations where water merges. Continue the work of filling water and building barriers until all the peaks are under water. Then the created barriers give the segmentation result.

Thresholding methods are applied to segment, not only grayscale images but also color images. For color images, one approach is to determine a separate threshold for each color channel of the image and then combine them with an AND operation. Segmentation based on color information (e.g., RGB, HSL, and HSV color models) may be more accurate than grayscale images [11, 12].

1.2. Edge-based segmentation

An edge is the boundary between two regions with different properties; it represents the change from one object or surface to another. Edges are used to characterize the physical extent of objects, since there is often a sharp adjustment in intensity at the region boundaries. Thus, detection of edges is a very important step toward image feature understanding; it is often used to divide images into areas corresponding to different objects.

The main idea of most edge-detection techniques is the computation of a local derivative of an image, including first- and second-order derivatives. The first-order derivative of choice in image processing is the gradient; it can be used to detect the presence of an edge in an image. Second-order derivatives in image processing generally are computed using the Laplacian. The sign of the second derivative can be used to determine whether an edge pixel is on the dark or light side of an edge [13–17]. Gradient operator and Laplacian operator are defined as follows:

(1) Gradient operators: the gradient of an image, $f(x, y)$, at location (x, y) is defined as the vector.

$$\nabla f(x, y) = \begin{bmatrix} g_x \\ g_y \end{bmatrix} = \begin{bmatrix} \frac{\partial f}{\partial x} \\ \frac{\partial f}{\partial y} \end{bmatrix} \tag{2}$$

The magnitude of this vector is computed from.

$$|\nabla f(x,y)| = \text{mag}(\nabla f(x,y)) = \left(g_x^2 + g_y^2\right)^{\frac{1}{2}} = \left[(\partial f/\partial x)^2 + (\partial f/\partial y)^2\right]^{\frac{1}{2}} \tag{3}$$

To simplify this computation, this quantity is approximated by using the following expression:

$$\nabla f \approx |g_x| + \left|g_y\right| \tag{4}$$

The direction angle of gradient vector is the maximum rate of change of f at coordinates (x,y):

$$\theta(x,y) = \arctan\left(\frac{g_y}{g_y}\right) \tag{5}$$

For a discrete image, the gradient can be calculated by the following expressions:

$$\begin{aligned} |\nabla f(x,y)| &= \left\{[f(x,y) - f(x+1,y)]^2 + [f(x,y) - f(x,y+1)]^2\right\}^{\frac{1}{2}} \\ &\approx |f(x,y) - f(x+1,y)| + |f(x,y) - f(x,y+1)| \end{aligned} \tag{6}$$

or

$$\begin{aligned} |\nabla f(x,y)| &= \left\{[f(x,y) - f(x+1,y+1)]^2 + [f(x+1,y) - f(x,y+1)]^2\right\}^{\frac{1}{2}} \\ &\approx |f(x,y) - f(x+1,y+1)| + |f(x+1,y) - f(x,y+1)| \end{aligned} \tag{7}$$

Edges can be detected with the help of first-order derivative type operators, as follows:

a. Sobel edge detector: using convolutions with row and column edge gradient masks, it is suitable to detect edges along the horizontal and vertical axis.

b. Robert detection: calculates the square root of the magnitude squared of the convolution with the row and column edge detectors; it is able to detect edges that run along the vertical axes of 45° and 135°.

c. Prewitt operator: detects edges by calculating the Prewitt compass gradient filters that return the result for the largest filter response.

d. Kirsch edge detector: performs convolution using eight filters that are applied to calculate gradient.

e. Frei-Chen edge operator: uses only the row and column filters.

The gradient method detects the edges by looking for the maximum and minimum in the first derivative of the image. However, these gradient operators tend to be sensitive to noise.

(2) Laplacian operator: the Laplacian of an image $f(x,y)$ is defined as

$$\nabla^2 f(x,y) = \frac{\partial^2 f(x,y)}{\partial x^2} + \frac{\partial^2 f(x,y)}{\partial y^2} \tag{8}$$

where $\frac{\partial^2 f(x,y)}{\partial x^2} = f(x+1,y) + f(x-1,y) - 2f(x,y)$, $\frac{\partial^2 f(x,y)}{\partial y^2} = f(x,y+1) + f(x,y-1) - 2f(x,y)$, so that

$$\nabla^2 f(x,y) = [f(x+1,y) + f(x-1,y) + f(x,y+1) + f(x,y-1)] - 4f(x,y) \tag{9}$$

The Laplacian is seldom used directly for edge detection because, as a second-order derivative, it is unacceptably sensitive to noise, its magnitude produces double edges, and it is unable to detect edge direction. In order to improve the effectiveness of edge detection, the following algorithms are proposed:

a. Laplacian of Gaussian (LoG): Gaussian filtering is combined with Laplacian to break down the image where the intensity varies, to detect the edges effectively [16]. However, this operator cannot find the orientation of edge because of using the Laplacian filter [17].

b. Canny edge detector (colored edge detectors): uses a multistage algorithm to detect a wide range of edges in images. First, Gaussian convolution filter is applied to smooth the images in order to reduce the noise. Second, a first-derivative operator (e.g., Sobel, Robert algorithm) is applied to the smoothed image in order to output a gradient magnitude image. Third, a process of nonmaximum suppression is applied to get rid of spurious response to edge detection in order to give a thin line in the output. Fourth, a double threshold is applied to determine potential edges. Last, tracking edges by hysteresis helps to finalize the detection by suppressing all the other edges. Canny edge detector performance is good; the only drawback is that it takes more time to compute and it is more complex [18, 19].

1.3. Region-based segmentation

An edge-based technique may attempt to find the object boundaries and then locate the object itself by filling them in; a region-based technique takes the opposite approach.

Region-based segmentation algorithms operate iteratively by grouping together neighboring pixels that have similar properties (such as gray level, texture, color, shape) and splitting groups of pixels that are dissimilar in value [20, 21]. There are a variety of approaches of region-based segmentation. These methods can be classified into two categories:

(1) Region growing method: This is the simplest region-based segmentation method. It is also classified as a pixel-based segmentation method since it involves the selection of initial seed points.

This approach to segmentation groups of pixels or subregions into larger regions is based on predefined criteria. First, a set of seed points is selected based on some user criterion (e.g., pixels in a certain grayscale range). Second, the regions are grown from these seed pixels to neighboring pixels, which are examined to ascertain if they should be added to the region, according to a region membership criterion (e.g., pixel intensity, grayscale texture, or color). Third, the second process is iterated on, in the same manner as general data clustering algorithms.

Region-growing-based techniques are better than the edge-based techniques in noisy images where edges are difficult to detect. However, it is computationally expensive.

(2) Split-and-merge method: This method consists of two steps: region splitting and region merging.

Region splitting starts with the whole image as a single region and subdivides it into subsidiary regions recursively while a condition of homogeneity is not satisfied.

Region merging is the opposite of region splitting and works as a way of avoiding over-segmentation. It starts with small regions and merges the regions that have similar characteristics (such as gray level, variance).

There are a great number of various approaches [22–26] excepting for the methods described above, also including the improvements of above methods [27–37]; for example, matching-based segmentation, clustering-based segmentation, fuzzy-inference-based segmentation, generalized PCA (principal component analysis)-based segmentation. Each segmentation method has its advantages and disadvantages. A universal algorithm of segmentation does not exist, as each type of image corresponds to a specific approach. Therefore, choice of technique depends on peculiar characteristic of individual problems. The emphasis of this paper lies on an improved method of scene image segmentation based on mathematical morphological operator-toggle operator.

This chapter is organized as follows: Section 1 presents an overview of methodologies and algorithms for image segmentation. A new proposed image segmentation method is then introduced in Section 2. In Section 3, the experimental results are analyzed to prove the validity of the proposed method. Finally, the chapter concludes in Section 4.

2. Scene image segmentation based on mathematical morphology

Signs and public notices are ubiquitous indoors and outdoors, and they are often used for route finding, finding public places and other locations. The texts in natural scene images contain important information. Therefore, text detection has attracted wide interest due to its usefulness in a variety of real-world applications, such as robots navigation, assisting visually impaired people, tourist navigation, enhancing safe vehicle driving, and so on [38, 39]. To date, a great number of algorithms have been proposed for detecting text on scene images or video [40–49]. However, most approaches proposed in the past research contribute to detect the text regions by analyzing the entire image. The image is segmented into text regions and non-text regions according to their features, respectively. The performance of these methods relies on the text detection algorithm and image complexity. Actually, scene text is usually presented on signboards. Because of the uniform color for the background of signboard, the ideal way for extracting text from scene images is to cut out the signboard regions first, and then detect text from the signboard regions. Thus, this chapter aims to propose an algorithm for segmenting a natural scene image into homogeneous regions. In our method, we first perform the image segmentation in order to detect homogeneous regions. Signboard regions are then detected with a simple criterion in order to remove the noise, such as trees and other non-signboard areas. In the following subsections, the proposed method is described and discussed in detail.

2.1. Image smoothing preprocessing

A natural scene image, I_{rgb}, is supposed to be a bitmap image based on the RGB (red-green-blue) color model. A smoothing process, edge preserving smoothing filter (EPSF) [20], is first applied to I_{rgb}.

The EPSF is applied independently to every pixel using different coefficients as shown in the following convolution mask:

$$\frac{1}{\sum_{i=1}^{8} c_i}\begin{bmatrix} c_1 & c_2 & c_3 \\ c_4 & 0 & c_5 \\ c_6 & c_7 & c_8 \end{bmatrix} \tag{10}$$

where c_i $(i = 1, 2, 3, \cdots, 8)$ are calculated using the following equation:

$$c_i = (1 - d_i)^p, \text{ where } p \geq 1 \tag{11}$$

where d_i $(i = 1, 2, 3, \cdots, 8)$ are the Manhattan color distances, which are extracted between the target pixel and the 8 neighboring pixels in a 3×3 window. That is,

$$d_i = \frac{\left|I_R^0 - I_R^i\right| + \left|I_G^0 - I_G^i\right| + \left|I_B^0 - I_B^i\right|}{3 \times 255} \tag{12}$$

where I_R^0, I_G^0, I_B^0 are the RGB color values of the target pixel, and I_R^i, I_G^i, I_B^i are the RGB color values of the ith neighboring pixel.

The filtering of the image is achieved by applying the convolution mask, Eq. (10), on each of the three color channels. Factor p in Eq. (11) scales exponentially the color differences; it controls the amount of blurring performed on the image. A fixed value $p = 13$ is used for all of our experiments because this results in very good performance. The target pixel of the convolution mask is set to zero to remove impulsive noise.

Finally, a smoothed image, I_{EPSF}, is obtained. I_{EPSF} is then converted to a grayscale image I_{gray} through the following equation:

$$\mathrm{I_{gray}}(x, y) = 0.2126 \cdot \mathrm{I_{EPSF}^R}(x, y) + 0.7152 \cdot \mathrm{I_{EPSF}^G}(x, y) + 0.0722 \cdot \mathrm{I_{EPSF}^B}(x, y) \tag{13}$$

where (x, y) is the coordinate of target pixel. $I^R_{EPSF}(x, y)$, $I^G_{EPSF}(x, y)$, and $I^B_{EPSF}(x, y)$ are the intensities for red, green, and blue, respectively, of the smoothed image I_{EPSF}.

2.2. Homogeneous region segmentation

A measure of region homogeneity is variance (i.e., regions with high homogeneity will have low variance). In this section, a mathematical morphological operator, Toggle Mapping (TM) [34], is introduced to segment a grayscale image into homogenous regions according to the pixel intensity. This is a simple way to segment a grayscale image into homogeneous regions based on a toggle operator. Such operator is defined as follows:

$$H(x, y) = \begin{cases} 1 & \textit{if } D(x, y) - E(x, y) \leq th \\ 0 & \textit{if } D(x, y) - E(x, y) > th \end{cases} \tag{14}$$

where H is a binary image taking two values, th is a threshold value, and D and E are the dilation image and erosion image of input image, respectively.

In order to meet the needs of the application, Dorini [35] and Fabrizio [36] have modified and improved this operator by adding new factors or weight coefficients. In their algorithms, the toggle operator is used one time for segmenting an image, so the values of thresholds and coefficients are fixed in their algorithms. However, for different images, the optimal values for thresholds and coefficients should be different. In order to overcome over-segmentation or under-segmentation, we propose a new algorithm for grayscale image segmentation. In our method, the toggle operator is applied iteratively on input image, and the value of threshold is changed in each iteration step. This is because, while applying Eq. (14) on a grayscale image, in the output image, the area of connected component will increase with the increase of threshold value. **Figure 1** shows an example for the increment of threshold value. Based on such feature, we propose an approach that tries to search for homogeneous regions by calculating the standard deviation of intensity for connected components. The detail procedure of our proposed algorithm is described in the following steps.

Step 1: Initializing $\leftarrow \phi$, $THSD = 20$, $ThRatio = 0.99$, $i \leftarrow 1$, $th \leftarrow i$, and applying Eq. (14) on I_{gray} then gets a set of connected components $CC^{(1)} = \left\{C^{(1)}_1, C^{(1)}_2, \cdots, C^{(1)}_{n(1)}\right\}$, where $n(1)$ is the number of elements of $CC^{(1)}$.

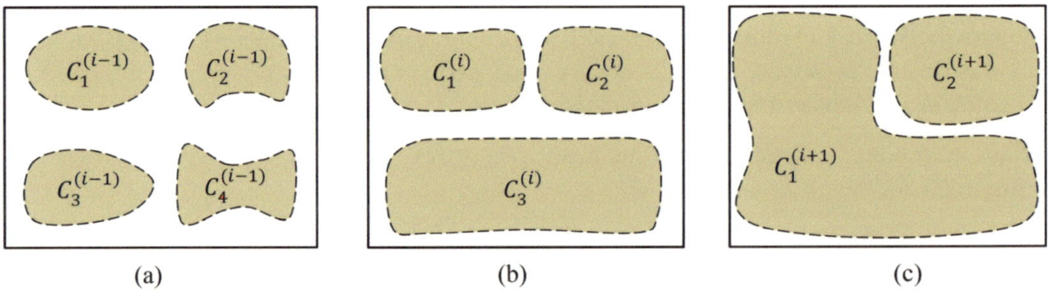

Figure 1. An example for result of Toggle Mapping (a) $th = i-1$, (b) $th = i$, (c) $th = i + 1$.

(a) (b) (c)

Figure 2. An example for natural scene image segmentation (a) Original Image, (b) Grayscale Image, (c) Labeling Image for Homogeneous Regions.

Step 2: Updating $i \leftarrow i + 1$, $th \leftarrow i$, and performing Eq. (14) on I_{gray} then gets a set of connected components $CC^{(i)} = \left\{C_1^{(i)},\ C_2^{(i)},\ \cdots,\ C_{n(i)}^{(i)}\right\}$, where $n(i)$ is the number of elements of $CC^{(i)}$. Setting $j \leftarrow 1$.

Step 3: For connected component $C_j^{(i)}$, where $C_j^{(i)} \in CC^{(i)}$, calculating its standard deviation of intensity $SD\_C_j^{(i)}$. If $SD\_C_j^{(i)} > THSD$, go to Step 4; else go to Step 5.

Step 4: Supposing $C_j^{(i)} \supseteq \left\{C_{j(1)}^{(i-1)}, C_{j(2)}^{(i-1)},\ \cdots, C_{j(m)}^{(i-1)}\right\}$, it means that $C_{j(1)}^{(i-1)},\ C_{j(2)}^{(i-1)},\ \cdots,\ C_{j(m)}^{(i-1)}$ are covered by $C_j^{(i)}$, where $\left\{C_{j(1)}^{(i-1)},\ C_{j(2)}^{(i-1)},\ \cdots,\ C_{j(m)}^{(i-1)}\right\}$ is a subset of $CC^{(i-1)}$. Updating $U \leftarrow U \cup \left\{C_{j(1)}^{(i-1)}, C_{j(2)}^{(i-1)},\ \cdots, C_{j(m)}^{(i-1)}\right\}$ and $CC^{(i)} \leftarrow CC^{(i)} - \left\{C_j^{(i)}\right\}$.

Step 5: Updating $j \leftarrow j + 1$. If $j \leq n(i)$, go to Step 3; else go to Step 6.

Step 6: Calculating the number of pixels $NP\_CC^{(i-1)}$ for $CC^{(i-1)}$ and $NP_CC^{(i)}$for $P\_CC^{(i)}$, respectively. If $\frac{\min\left(NP\_CC^{(i-1)},\ NP\_CC^{(i)}\right)}{\max\left(NP\_CC^{(i-1)},\ NP\_CC^{(i)}\right)} > ThRatio$, go to Step 7; else go to Step 2.

Step 7: Terminating. Finally, U is a set of homogeneous regions.

As shown in **Figure 2**, the natural scene image can be segmented into homogeneous regions. The result showed that our proposed method can work effectively with high accuracy.

3. Experiment and results

3.1. Experimental images

In our experiment, 500 natural scene images are captured with various signboards, shop names, traffic signs, and more. All the original images are saved in RGB24 bitmap format with a size of 1000 × 1500 pixels. In order to provide a wide range of real-life scenarios, images are captured with different compact digital cameras at different angles, positions, and under variable lighting and weather conditions. **Figure 3** shows some examples used in this experiment. **Table 1** shows the experimental environment.

3.2. Evaluation of image segmentation

In this subsection, our proposed method is compared to the methods of watershed segmentation using gradient [1], Canny edge detection [18], and region growing [1]. In order to evaluate the accuracy of our proposed method. There are many parameters included in not only our method but also the other three methods. Therefore, 100 images, selected from the total 500 images, are used for training and deciding the value of parameters based on the grid search method. The remaining 400 images that differ from the 100 training images are used to do the experiment in order to evaluate the accurancy of segmentation.

Figure 3. Examples of scene images.

OS	Microsoft Windows 10 Enterprise
CPU	Intel(R) Xeon(R) E5–2620 v4 2.10GHz (dual processor)
Memory	128GB
Programming Language	MATLAB

Table 1. Experimental environment.

The purpose of our research is to support visually impaired people to access the scene text. This paper aims to segment the natural scene images into homogenous regions. This is because, after the segmentation, specified criteria can be applied to select the signboard regions and the text can then be extracted from these regions. Therefore, in the experiment, we only focus on the accuracy of signboard segmentation.

The result of segmentation relies on not only segmentation algorithms but also the quality of the images. From the observation, the signboard regions can be segmented into four categories: PERFECT, FRAGMENT, EXCALATION, and FRAGMENT and EXCALATION.

PERFECT: a signboard has been segmented correctly, as shown in **Figure 4(a)**.

FRAGMENT: a signboard is segmented out in fragments, as shown in **Figure 4(b)**, where the extracted results are part of one signboard.

EXCALATION: one part of the signboard is extracted, but the others are lost, as shown in **Figure 4(c)**, where the extracted region is part of one signboard.

EXCALATION and FRAGMENT: one part of a signboard is segmented into fragments, but the other part is lost, as shown in **Figure 4(d)**.

In the experiment, PERFECT and FRAGMENT are evaluated as correct results, EXCALATION and EXCALATION and FRAGMENT are evaluated as incorrect results.

There are 482 target signboards in 400 experimental images. After the experiment, the accuracy of signboard segmentation and the average image processing time are calculated, respectively. The results are shown in **Table 2**.

As shown in **Table 2**, the average processing time of our proposed method is not so short. This is because our algorithm iteratively applies the toggle operator to segment image and find homogeneous regions. So, it is time-consuming. For the region-growing method, it first searches the seeds in an image and then performs the growing processing. This is also time-consuming.

Figure 4. Cases of signboard segmentation (a) Complete Signboard, (b) Fragmentary Signboard, (c) Partial Signboard, (d) Partially & Fragmentary.

Methods	Accuracy (%)	Average execution time (s)
Our method	94.5	3.21
Watershed segmentation	83.6	2.82
Canny edge detector	91.4	0.73
Region growing	92.2	4.08

Table 2. Segmentation accuracy and average processing time.

Figures 5–8 are the segmentation results of **Figure 3**, by applying our menthod, watershed segmentation, Canny edge detection, and region-growing method, respectively. From the observation, our proposed algorithm can segment an image into homogeneous regions effectively, and some results are better than those applying the Canny edge detector and the region-growing method, because of the Canny edge operator not always detecting the closed boundary of object and the result of region-growing method deeply depending on the initial seeds selection.

Each method can achieve a high accuracy value if the quality of images is very good. But if the images include much noise, the accuracy of segmentation is very low any method. The signboard regions cannot be segmented completely due to the following reasons: (1) the surface of signboard is corroded, for example, **Figure 9(a)**; (2) shadow exists on the signboard, for example, **Figure 9(b)**; and (3) reflective effect, for example, **Figure 9(c)**.

Figure 5. Result of our method for segmenting **Figure 3** (a) Homogeneous Regions, (b) Homogeneous Regions, (c) Homogeneous Regions, (d) Homogeneous Regions, (e) Homogeneous Regions, (f) Homogeneous Regions.

http://dx.doi.org/10.5772/intechopen.72603

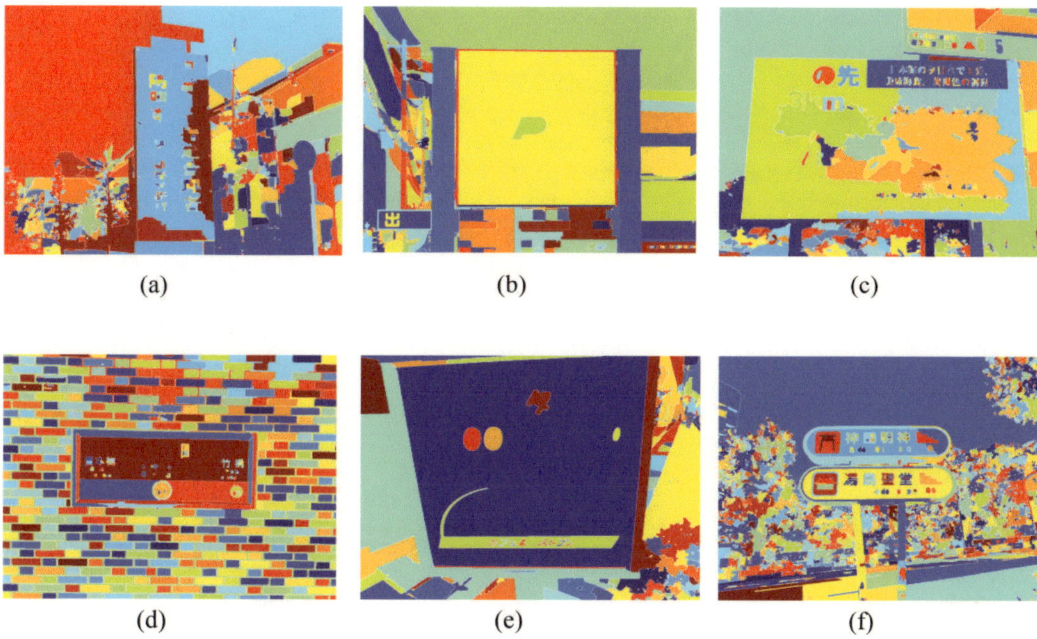

Figure 6. Watershed segmentation result of **Figure 3** (a) Homogeneous Regions, (b) Homogeneous Regions, (c) Homogeneous Regions, (d) Homogeneous Regions, (e) Homogeneous Regions, (f) Homogeneous Regions.

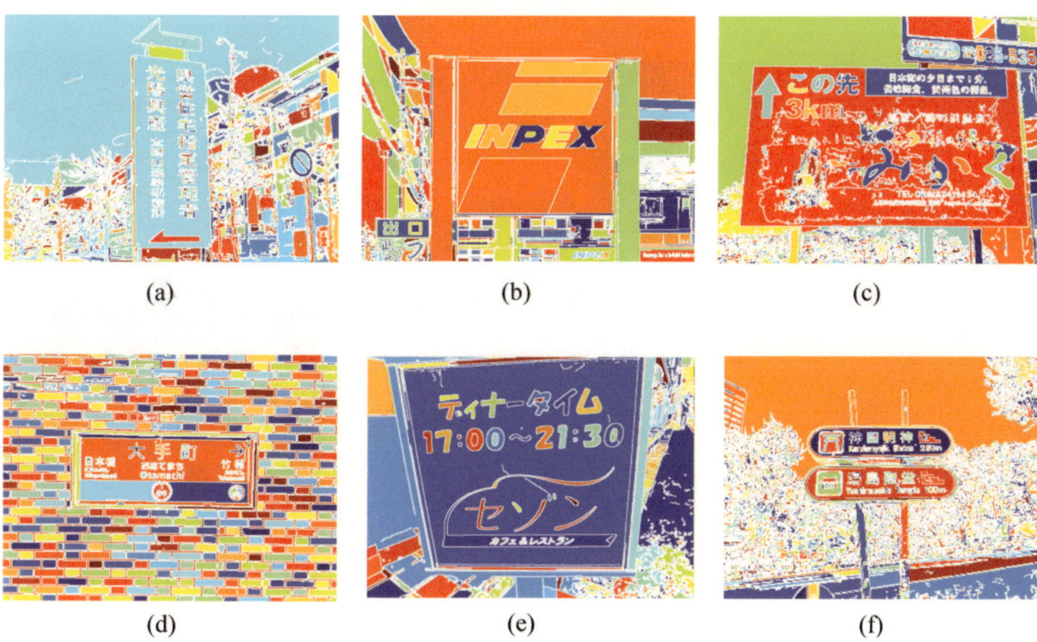

Figure 7. Canny edge detector for segmentation of **Figure 3** (a) Homogeneous Regions, (b) Homogeneous Regions, (c) Homogeneous Regions, (d) Homogeneous Regions, (e) Homogeneous Regions, (f) Homogeneous Regions.

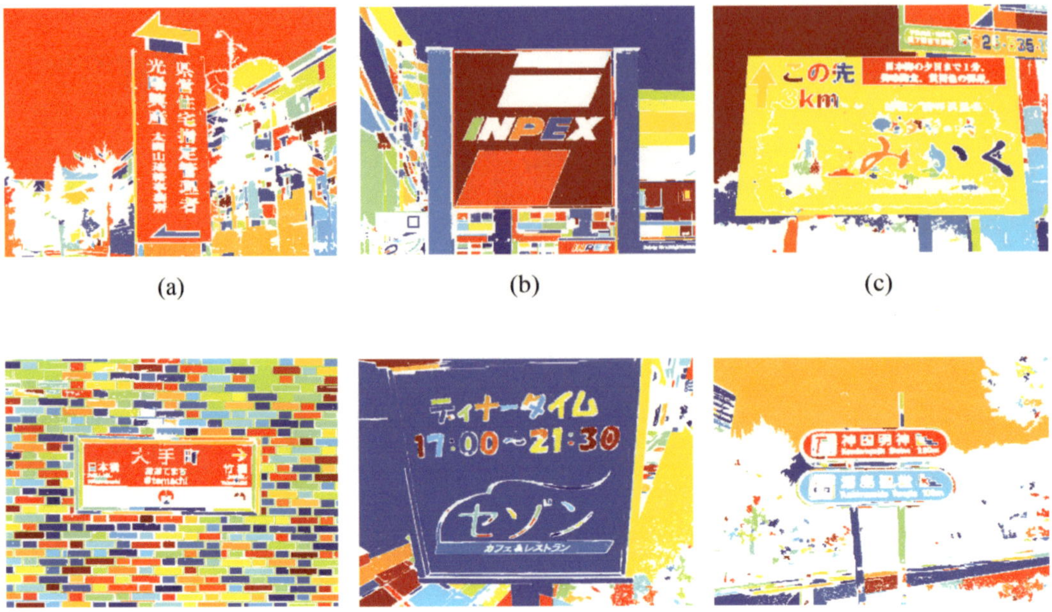

(a) (b) (c)

(d) (e) (f)

Figure 8. Region growing for segmentation of **Figure 3** (a) Homogeneous Regions, (b) Homogeneous Regions, (c) Homogeneous Regions, (d) Homogeneous Regions, (e) Homogeneous Regions, (f) Homogeneous Regions.

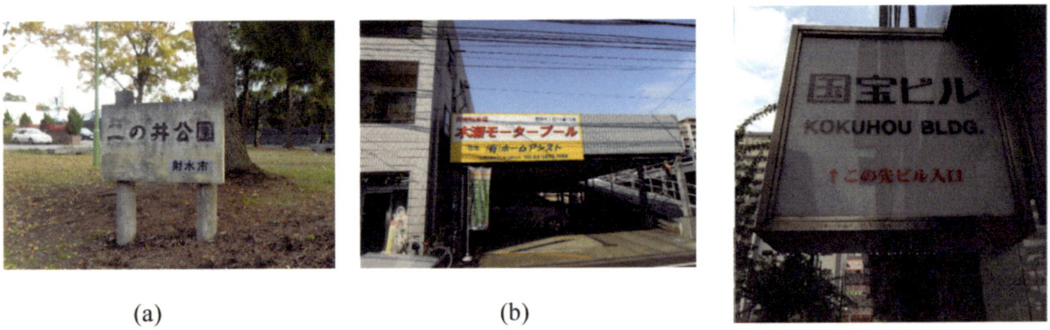

(a) (b) (c)

Figure 9. Examples of scene images.

4. Summary

This chapter proposes a method of mathematical morphology-based natural scene image segmentation. First, a number of typical segmentation algorithms are reviewed and discussed, and the objective of this chapter was introduced. Second, our proposed method was described. Third, the experiment was done and discussed.

The proposed method was tested on different images, and the results showed that our method can be an effective way for scene images segmentation. However, the results indicated that the signboard regions were extracted with low accuracy due to the presence of shadows or even corroded signboards.

In order to improve the accuracy of segmentation results, in the near future, we will introduce techniques for removing shadows and reflections in images.

Acknowledgements

This work is supported by the Natural Science Foundation of the Higher Education Institutions of Jiangsu Province, China (Grant No. 17KJB520007, 17KJB470002); Doctoral Research Foundation of Jiangsu University of Science and Technology, China (Grant No. 1624821607–9); and Natural Science Foundation of Jiangsu Province, China (Grant No. BK20150471).

Author details

Jianjun Chen[1,2]*, Haijian Shao[1] and Chunlong Hu[1]

*Address all correspondence to: kengunchin@foxmail.com

1 Jiangsu University of Science and Technology, Zhenjiang, China

2 Toyama Prefectural University, Imizu, Japan

References

[1] Gonzalez RC, Woods RE. Digital Image Processing. (3rd edition), Pearson International 3rd ed. New Jersey, NJ: Pearson Prentice Hall; 2008

[2] Shapiro LG, Stockman GC. Computer Vision. New Jersey: Prentice-Hall; 2001. pp. 279-325

[3] Zhang YJ, Gerbrands JJ. Objective and quantitative segmentation evaluation and comparison. Signal Processing. 1994;**39**(1–2):43-54

[4] Zhang YJ. A survey on evaluation methods for image segmentation. Pattern Recognition. 1996;**29**(8):1335-1346

[5] Sezgin M, Sankur B. Survey over image thresholding techniques and quantitative performance evaluation. Journal of Electronic Imaging. 2004;**13**(1):146-165

[6] Tobias OJ, Seara R. Image segmentation by histogram thresholding using fuzzy sets. IEEE Transactions on Image Processing. 2002;**11**(12):1457-1465

[7] Luessi M, Eichmann M, Schuster GM, Katsaggelos AK. Framework for efficient optimal multilevel image thresholding. Journal of Electronic Imaging. 2009;**18**(1):013004-1-013004-10

[8] Lai YK, Rosin PL. Efficient circular thresholding. IEEE Transaction on Image Processing. 2014;**23**(3):992-1001

[9] Saini R, Dutta M, Kumar R. A comparative study of several image segmentation techniques. Journal of Information and Operations Management. 2012;**3**(1):21-24

[10] Otsu N. A threshold selection method from gray-level histograms. IEEE Transactions on Systems, Man, and Cybernetics. 1979;**9**(1):62-66

[11] Pham N, Morrison A, Schwock J, et al. Quantitative image analysis of Immunohistochemical stains using a CMYK color model. Diagnostic Pathology. 2007:2-8

[12] Plataniotis KN, Venetsanopoulos AN. Color Image Processing and Applications. Berlin, Heidelberg: Springer; 2000

[13] Senthilkumaran N, Rajesh R. Edge detection techniques for image segmentation—A survey of soft computing approaches. International Journal of Recent Trends in Engineering. 2009;**1**(2):250-254

[14] Senthilkumaran N Rajesh R Edge detection techniques for image segmentation—A surveyIn: Proceedings of the International Conference on Managing Next Generation Software Applications. 2008. pp. 749–760

[15] Torre V, Poggio T. On edge detection. IEEE Transaction Pattern Analysis Machine Intelligence. 1986;**8**(4):147-163

[16] Acharjya PP, Das R, Ghoshal D. A study on image edge detection using the gradients. International Journal of Scientific and Research Publications. 2012;**2**(12):1-5

[17] Yogamangalam R, Karthikeyan B. Segmentation techniques comparison in image processing. International Journal of Engineering and Technology. 2013;**5**(1):307-313

[18] Canny J. A computational approach to edge detection. IEEE Transactions on Pattern Analysis and Machine Intelligence. 1986;**PAMI-8**(6):679-698

[19] Lakshmi S, Sankaranarayanan V. A study of edge detection techniques for image segmentation computing approaches. Special Issue on Computer Aided Soft Computing Techniques for Image and Biomedical Applications. 2010:35-41

[20] Nikolaou N, Papamarkos N. Color reduction for complex document images. International Journal of Imaging Systems & Technology. 2009;**19**:14-269

[21] Manjula KA. Role of image segmentation in digital image processing for information processing. International Journal of Computer Science Trends and Technology. 2015;**3**(3): 312-318

[22] Liu D, Ma L, Chen H, Meng K Medical image segmentation based on improved fuzzy C-means clustering In: Proceedings of the International Conference on Smart Grid and Electrical Automation. 2017. pp. 406-410

[23] Fabijańska A Corneal endothelium image segmentation using feedforward neural networkIn: Proceedings of the Federated Conference on Computer Science and Information Systems; 2017. pp. 629-637

[24] Liu Y, Cao F, Zhao J, Chu J. Segmentation of white blood cells image using adaptive location and iteration. IEEE Journal of Biomedical and Health Informatics. 2017;**21**(6): 1644-1165

[25] Liu W, Qing X, Zhou J. A novel image segmentation algorithm based on visual saliency detection and integrated feature extraction. In: Proceedings of the International Conference on Communication and Electronics Systems; 2016. pp. 1-5

[26] Swarnalatha A, Manikandan M Review of segmentation techniques for intravascular ultrasound (IVUS) images In: Proceedings of the Fourth International Conference on Signal Processing, Communication and Networking. 2017. pp. 1-4

[27] Wangchamhan T, Chiewchanwattana S, Sunat K. Multilevel thresholding selection based on chaotic multi-verse optimization for image segmentation. In: Proceedings of the 13th International Joint Conference on Computer Science and Software Engineering; 2016. pp. 1-6

[28] Zhou X, Li Y, Shen L. A novel adaptive local thresholding approach for segmentation of HEp-2 cell images. In: Proceedings of the IEEE International Conference on Signal and Image Processing; 2016. pp. 174-178

[29] Mirah Kasturi M, Sutha P, Jayanthi VE. efficient edge detection method for diagnosis of 2D and 3D lung and liver images. In: Proceedings of the International Conference on Inventive Systems and Control; 2017. pp. 1-5

[30] Hien NM, Binh NT, Viet NQ Edge detection based on fuzzy c means in medical image processing systemn: Proceedings of the International Conference on System Science and Engineering; 2017. pp. 12-15

[31] Günen MA, Atasever ÜH, Beşdok E. A novel edge detection approach based on backtracking search optimization algorithm (BSA) clustering. In: Proceedings of the 8th International Conference on Information Technology; 2017. pp. 116-122

[32] Albano JA, Ziemann AK, Messinger DW. Assessing the impact of the edge-weighting function in a graph-based approach to anomaly detection. In: Proceedings of the 5th Workshop on Hyperspectral Image and Signal Processing: Evolution in Remote Sensing; 2013. pp. 1-4

[33] Bi C, Yuan Y, Zhang R, Xiang Y, Wang Y, Zhang J. A dynamic mode decomposition based edge detection method for art images. IEEE Photonics Journal. 2017;**99**:1-1

[34] Liu Y, Cheng M, Hu X, Wang K, Bai X. Richer convolutional features for edge detection. In: Proceedings of the IEEE Conference on Computer Vision and Pattern Recognition; 2017. pp. 5872-5881

[35] Manisha, Radhakrishnan B, Padma Suresh L. Tumor region extraction using edge detection method in brain MRI images. In: Proceedings of the International Conference on Circuit, Power and Computing Technologies; 2017. pp. 1-5

[36] Grinias, Tziritas G. Region-level moving object segmentation by graph labeling. In: Proceedings of the 13th European Signal Processing Conference; 2005. pp. 1-4

[37] Liu Z, Wu X, Zhao B, Peng Q. Clothing extraction using region-based segmentation and pixel-level refinement In: Proceedings of the IEEE International Symposium on Multimedia 2014. pp. 303-310

[38] Wang S, Fu C, Li Q. Text detection in natural scene image: A survey. In: Proceedings of the International Conference on Machine Learning and Intelligent Communications; 2017. pp. 257-264

[39] Liang J, Doermann D, Li H. Camera-based analysis of text and documents: A survey. International Journal of Document Analysis and Recognition. 2005;**7**(2–3):84-104

[40] Matsuda Y, Omachi S, Aso H. String detection from scene images by binarization and edge detection. Institute of Electronics, Information and Communication Engineers, D. 2010;**J93-D**(3):336-344

[41] Hase H, Yoneda M, Sakai M, Maruyama H. Consideration of color segmentation to extract character areas from color document images. Institute of Electronics, Information and Communication Engineers, DII. 2000;**J83-D-II**(5):1294-1304

[42] Ashida K, Nagai H, Okamoto M, Miyao H, Yamamoto H. Extraction of characters from scene images. Institute of Electronics, Information and Communication Engineers, DII. 2005;**J88-D-II**(9):1817-1824

[43] Saitoh S, Goto H, Kobayashi H. Analysis and comparison of frequency features for scene text detection. In: Technical Report of IEICE, PRMU2004-128. 2004. pp. 31-36

[44] Crandall D, Antani S, Kasturi R. Extraction of special effects caption text events from digital video. International Journal on Document Analysis and Recognition. 2003;**5**:138-157

[45] Kalai Selvi U, Anish Kumar J. Camera based assistive text reading system using gradient and stroke orientation for blind person. International Journal of Latest Trends in Engineering and Technology. 2014;**4**(1):325-330

[46] Yao C, Bai X, Liu W, et al. Detecting texts of arbitrary orientations in natural images. In: Proceedings of the IEEE Conference on Computer Vision & Pattern Recognition; 2012. pp. 1083-1090

[47] Yin XC, Yin X, Huang K, et al. Robust text detection in natural scene images. IEEE Transaction on Pattern Analysis and Machine Intelligence. 2014;**36**(5):970-983

[48] Shivakumara P, Phan TQ, Tan CL. New Fourier-statistical features in RGB space for video text detection. IEEE Transactions on Circuits and Systems for Video Technology. 2010;
20(11):1520-1532

[49] Neumann L, Matas J. A method for text localization and recognition in real-world images. In: Proceedings of the Asian Conference on Computer Vision; 2010. pp. 770-783

Chapter 3

Color Reconstruction and Resolution Enhancement Using Super-Resolution

Eduardo Quevedo Gutiérrez and
Gustavo Marrero Callicó

Additional information is available at the end of the chapter

http://dx.doi.org/10.5772/intechopen.71262

Abstract

Image super-resolution (SR) is a process that enhances the resolution of an image or a set of images beyond the resolution of the imaging sensor. Although there are several super-resolution methods, fusion super-resolution techniques are well suited for real-time implementations. In fusion super-resolution, the high-resolution images are reconstructed using different low-resolution-observed images, thereby increasing the high-frequency information and decreasing the degradation caused by the low-resolution sampling process. In terms of color reconstruction, standard reconstruction algorithms usually perform a bilinear interpolation of each color. This reconstruction performs a strong low-pass filtering, removing most of the aliasing present in the luminance signal. In this chapter, a novel way of color reconstruction is presented by using super-resolution in order to reconstruct the missing colors.

Keywords: super-resolution, color reconstruction, video enhancement, image fusion, resolution enhancement

1. Introduction

The technical limitations of the imaging devices clearly restrict the spatial resolution of video sequences. The super-resolution (SR) reconstruction technique is usually defined in the state of the art of image processing as a method that combines multiple low-resolution (LR) images with some amount of aliasing to obtain a higher resolution image. Although several methods have been implemented in this field, there are still several future open research challenges [1]. SR can be found in fields like surveillance, remote sensing, astronomy, and an extensive set of consumer electronics applications, among many others [2–5].

This chapter proposes a combination of color reconstruction and resolution enhancement, showing how to apply SR as a novel way to reconstruct missing colors. The quality improvement applying SR mainly depends on two factors: movement in the sequence and presence of aliasing in the image [6, 7]. Therefore, it is important to assure some amount of aliasing in the image without using any artificial system. This aliasing is a direct consequence of the acquisition system.

This chapter is organized as follows: the super-resolution concept and the steps of the considered algorithm are introduced in Section 2, the experimental methodology is presented in Section 3, and the dataset is described in Section 4, while Section 5 shows some color reconstruction algorithms, and Section 6 shows the color reconstruction using super-resolution. Section 7 presents a novel combination between color reconstruction and resolution enhancement. Finally, the conclusions are highlighted in Section 8.

2. Super-resolution

The first frequency domain-based SR algorithm was proposed by Huang and Tsay in 1984 [8], and, after that, several other SR techniques have been developed with different results. A general consideration governing this kind of algorithms is that in order to obtain significant improvements in the resulting image, some amount of aliasing in the input LR images must be guaranteed. This approach is called fusion super-resolution in the literature. The main advantages are that it requires limited processing and memory resources [9] but has the disadvantage of generating artifacts when applied to real-life video sequences.

In order to avoid artifacts, the authors have proposed several improvements based on selective filters and the usage of several cameras [6, 7]. The result consists on a non-iterative dynamic SR algorithm, with reduced computational and memory requirements. It provides static and dynamic SR depending on whether the result is one single frame and a video sequence, respectively. The SR algorithm used in this work has been extensively tested by the authors for general-purpose images and video sequences [9, 10] and also in specific applications, like underwater images [11] or compressed images [12]. As this SR algorithm is based on the direct processing of the information contained in the input images and not in machine learning techniques, it can be used for any kind of situations (including local and global motion, changes in the illumination among frames, etc.), and it does not depend on any training set of images.

2.1. Algorithm description

The SR algorithm considered in this chapter starts after the adjustment of the borders of the images to normalize their dimensions. An LR image sequence is the input for the motion estimation, in which the goal is to find the best motion vectors (MVs) that estimate the real movement (with sub-pixel accuracy) between two consecutive frames. The entire algorithm is composed of a set of successive steps, the first of which is the motion estimation and the second one is the motion compensation, in this case called Shift & Add (S&A), which uses the MVs to build a grid for the frame obtained fusing the LR images. The Holes Filling phase substitutes the zeros with an interpolation of the values coming from adjacent pixels. Since usually big images are considered in the sequence (with the consequent high computational load), the algorithm introduces the concept of window-selective filter (WSF), based on a working window (WW). The aim is to apply SR on the current frame (CF) using only the pictures comprised in the same WW (see **Figure 1**).

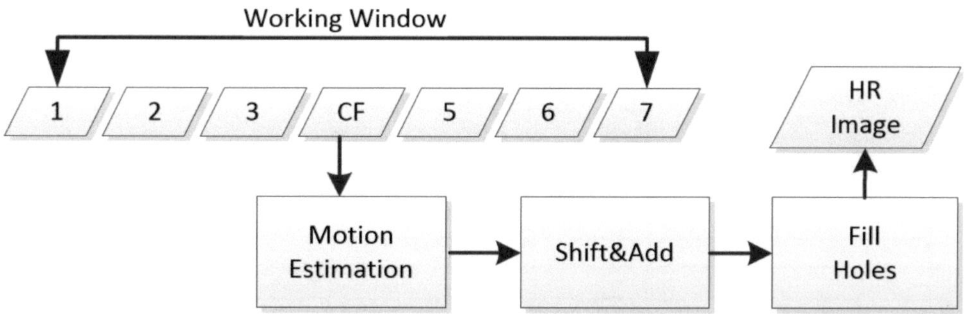

Figure 1. Algorithm scheme that shows the working window (in this case frames between 2 and 6), the current frame (CF), and the other images (always taking into consideration only the Y component). The CF undergoes motion estimation and after that the Shift & Add, followed by the Holes Filling process.

2.2. Motion estimation

In the motion estimation (ME) stage, the image is divided into non-overlapping macro-blocks (MBs), following the block-matching principle (a very popular and efficient ME technique, especially in video compression). In this way, not only global but also local movements are evaluated, referred to the CF. In fact, taking into consideration only the global motion evaluation, the algorithm will fail to create high-resolution (HR) frames when there is a considerable amount of local movement between frames.

2.3. Shift & Add

During this phase, the frames of the WW are used, on a block-per-block basis, to form the final HR image. Firstly, a grid of four times the dimension of the initial LR image is created: it contains the image of the CF after the ME, and the positions (in fact high-resolution pixels) in which no information is present are filled with zeros. MBs of the CF are considered as LR macro-blocks, and every one of them is interpolated to create the base for the corresponding HR block. Only the zero values are interpolated in this process as they represent empty positions.

2.4. Holes Filling

At the MB level, MVs on both horizontal and vertical axes are checked to be at the sub-pixel level, and the sum of absolute difference (SAD) parameter is used as a weight function between MVs and MBs. If new information is detected, it is added to the grid, whereas the points that remain empty are filled with interpolations of adjacent values, a procedure that goes under the name of Holes Filling process.

3. Experimental methodology

The experimental methodology is based on a modification on the capturing system. A common capture system is shown in **Figure 2**. In front of the system, there is a lens to focus the

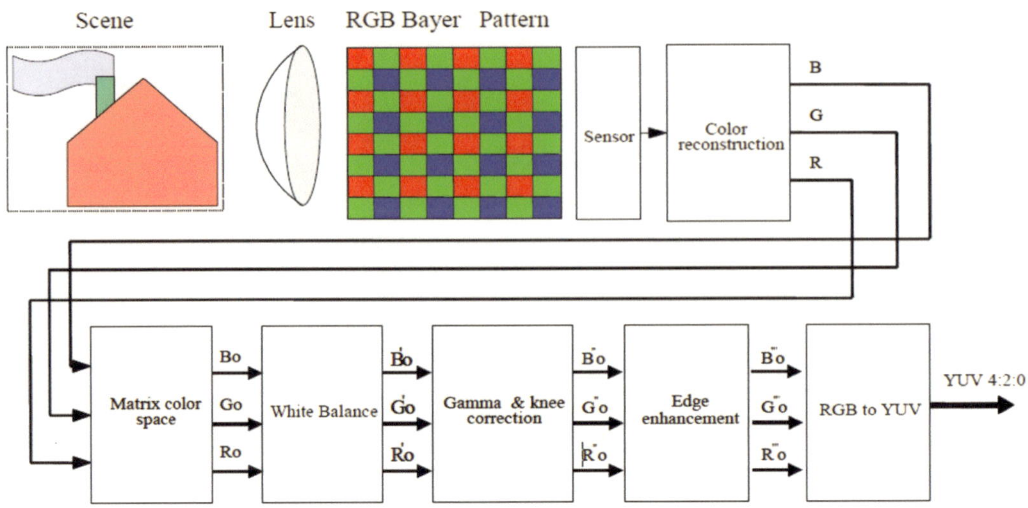

Figure 2. Capture system for imaging.

scene on the sensor plane. This plane is commonly called image plane. The most common technique to obtain a color image is to sense red, green, and blue colors. In the high-quality products, three sensors are placed, one for each of these colors. Using some mechanical or optical system, the image plane is focused on the three sensors, obtaining three colors at the full sensor resolution.

In cheaper applications, only one sensor is used. In that case, in order to obtain three colors, a color filter array (CFA) is placed in front of the sensor. By this way, each sensor cell senses a different color. The most commonly used is the red, green, and blue (RGB) Bayer pattern, which is represented at the beginning of **Figure 2**.

After sensing the colors, a color reconstruction algorithm is applied to obtain the colors where they were not sensed. There are several color reconstruction algorithms; most of them are based on different types of interpolation [13–16]. After the color reconstruction, another image processing is performed to obtain the image in YUV format, which is the input of the compression system.

In order to obtain the YUV 4:2:0 image, the following operations are done:

- Calculate the YUV 4:4:4 image as:
  - $Y = 0.3 \cdot R + 0.59 \cdot G + 0.11 \cdot B$
  - $U = Y - R$
  - $V = Y - B$
- Perform a low-pass filtering on U and V images.
- Subsample the U and V images to obtain the YUV 4:2:0 format.

In order to keep aliasing in this work, three aspects are of special relevance:

- Lens
- Sensor configuration
- Color reconstruction algorithm

In this study, some sequences were recorded using high-quality lens. The resolution of this lens was selected to be higher than the sensor resolution. By this way, we avoided the lens optical low-pass filtering (OLPF). In this case, the aliasing appears in the sampling process. This process is guided by two main factors:

- The sensor fill factor
- The color sampling

The sensor fill factor describes the ratio between the light-sensitive area per pixel and the total pixel area. This factor can be increased by using small micro-lens at the top of each pixel. With these lenses, shown in **Figure 3**, some of the incident light in the nonsensitive area is concentrated to the light-sensitive area. Those micro-lenses can increase the fill factor up to 60 or 70% but remove some amount of aliasing due to the optical low-pass filter (OLPF) effect produced by the lenses. For this work, we selected a sensor without micro-lenses, thus keeping the aliasing that in other cases would be removed by the OLPF.

The second factor is the color sampling. In fact, each color image is a reconstructed image. The maximum sampled frequency is represented in **Figure 4**. The axes of this figure are the horizontal and vertical space frequencies, with the low frequencies located in the middle of the square. The color squares represent the maximum sampled frequency for each color: the bigger square with a diamond shape represents the frequencies of the green color, and the smaller inner squares represent the frequencies of the red and blue colors. It is important to note that the green signal has higher frequencies than red and blue ones. This limitation in the frequency distribution produces some aliasing in the color spectral domain. In this work a CFA has been used, largely decreasing the resolution of the three color components, but in special the resolution of the red and blue components.

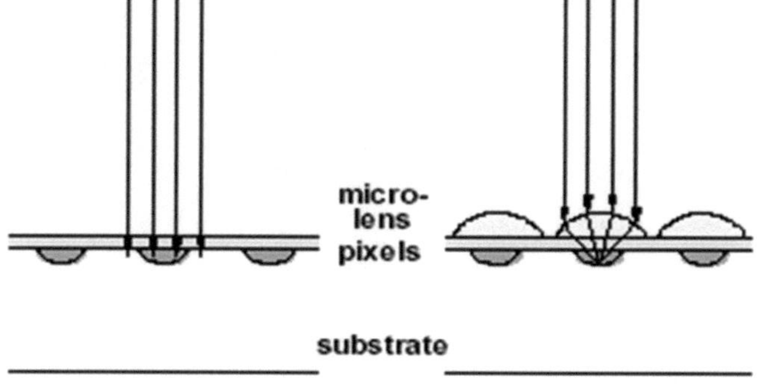

Figure 3. Micro-lenses disposition.

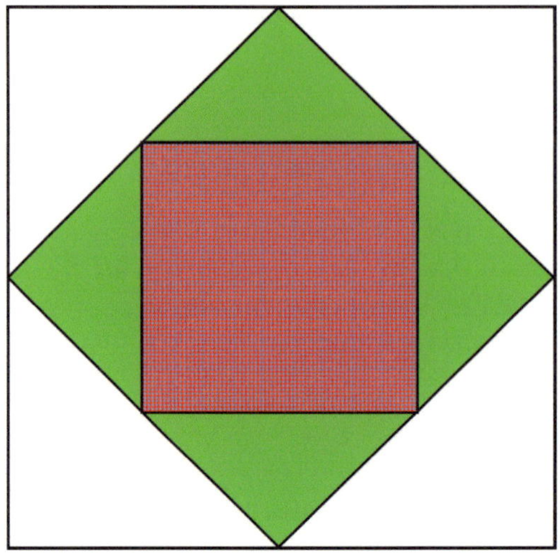

Figure 4. Color-sampled frequencies in the frequency space.

4. Description of the dataset

Following the methodology described in Section 3, a set of 20 images with random sub-pixel shifts among them was obtained from a Dutch newspaper. The original image exhibits a rich amount of color details and represents a picture of the Big Ben cleaners working on the clock on the left of the image and two columns of text on the right of the image. The sizes of all the sampled images are 624 pixels wide and 464 pixels in height. Each sample image has three color components, with a bit depth of 8 bits per color component (24 bits/pixel). The first LR input image can be seen in **Figure 5**.

Figure 5. Frame 1 of the input image used to reconstruct the color.

http://dx.doi.org/10.5772/intechopen.71262

5. Color reconstruction algorithms

In this section, a first experiment consisting in the use of the image in YUV 4:2:0 for resolution enhancement is presented. In this case, the image follows all the processing, but different algorithms to reconstruct the color signal are chosen. This processing must keep as much aliasing as possible in the luminance image, to obtain a good performance in SR. Three algorithms were tested: standard reconstruction, SmartGreen1, and SmartGreen3 [17].

5.1. Standard reconstruction algorithms

Standard reconstruction algorithms perform a bilinear interpolation of each color plane. This reconstruction performs a strong low-pass filtering, removing most of the aliasing present in the luminance signal. Super-resolution algorithm was applied using the obtained image in YUV 4:2:0. The results are shown in **Figures 6** and **7**. After 20 frames, only a small improvement was achieved due to the limited aliasing present in the luminance image.

5.2. SmartGreen1

SmartGreen1 [17] performs a bilinear interpolation over the red and blue images. If high frequencies are present, the green image is obtained from the red or blue images in places where there are no green samples. In those places, the green image is equal to the red- or blue-sampled image multiplied by some predefined coefficients.

If this approach works well, the green image can improve the pixel resolution and, therefore, increases the luminance resolution. This algorithm works better in gray areas, where R = G = B. SmartGreen1 seems to keep more aliasing information, and therefore we expected better performance when used in combination with super-resolution. Results are shown in **Figures 8** and **9**.

Figure 6. Standard reconstruction with SR of "paper," frame 1 (left) and frame 20 (right). For the first frame, no SR is possible, and so it is equal to bilinear interpolation.

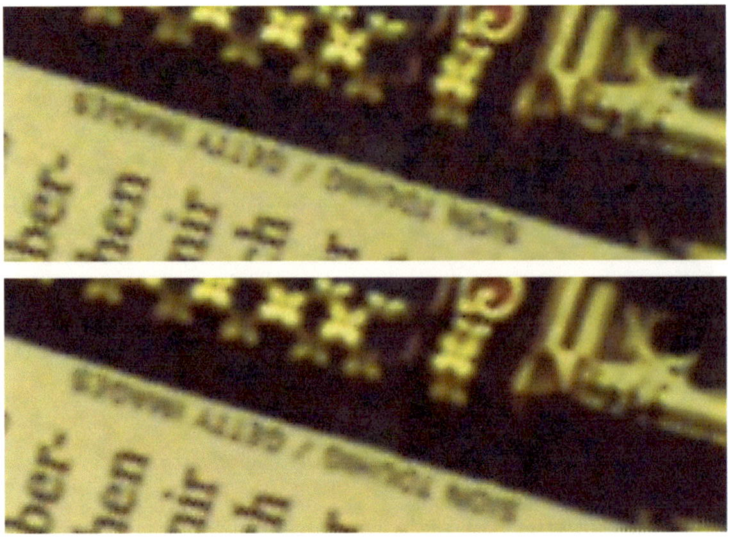

Figure 7. Standard reconstruction with SR of "paper," frame 1 (up) and frame 20 (down) details.

Figure 8. SmartGreen1 reconstruction with SR of "paper," frame 1 (left) and frame 20 (right). For the first frame, no SR is possible, and so it is equal to bilinear interpolation.

5.3. SmartGreen3

SmartGreen3 requires more computational load. In previous versions of SmartGreen, the computed green image from the red and blue values was used directly. This approach worked well in gray areas but made colorful images grayer. In SmartGreen3, this image is only used to enhance the edges and to help in the detection of false color due to color aliasing. The main issue with this algorithm is that it performs a band-pass filtering of the luminance image, removing aliasing. In consequence, we will not increase the quality of the image by using super-resolution. Due to the good performance of this technique, it can be a reference

http://dx.doi.org/10.5772/intechopen.71262

Figure 9. SmartGreen1 reconstruction with SR of "paper," frame 1 (up) and frame 20 (down) details.

of sharpness for SR as one of the best qualities that the color reconstruction algorithms can achieve. The results obtained using these algorithms are shown in **Figures 10** and **11**.

The performance obtained using standard reconstruction or SmartGreen1 with SR is not better than using SmartGreen3. Moreover, the improvement of using SR in conjunction with SmartGreen3 is very small, so it is not very wise to apply SR in that case due to the computational load increase.

Figure 10. SmartGreen3 reconstruction with SR of "paper," frame 1 (left) and frame 20 (right). For the first frame, no SR is possible, and so it is equal to bilinear interpolation.

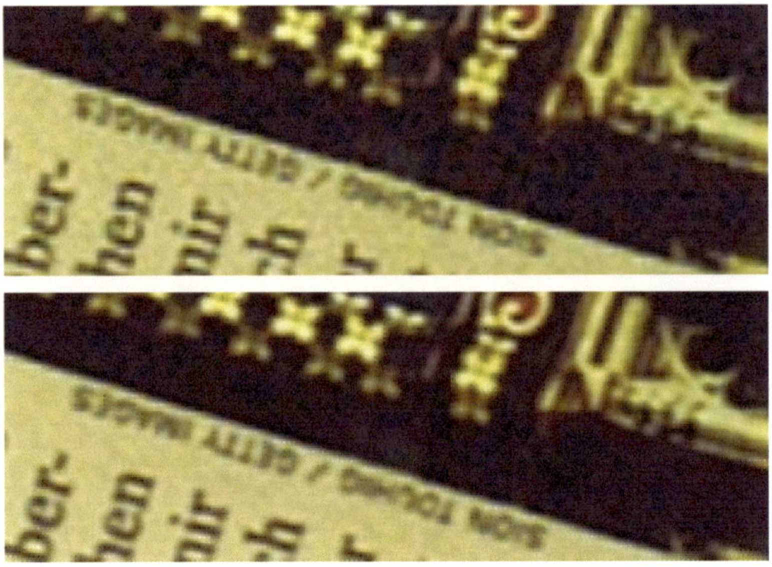

Figure 11. SmartGreen3 reconstruction with SR of "paper," frame 1 (up) and frame 20 (down) details.

6. Color reconstruction using super-resolution

As the results obtained were not very satisfactory using the previous approaches, a new approach was tried. Since each color image is subsampled, a large amount of aliasing is present in each one of these images. In fact, when using the Bayer pattern, the luminance image is never fully sensed because it is composed by sampled values of one color and reconstructed values from the other two color images. Because the sensor works in RGB, it is better to work directly in this domain.

Following this idea, it can be seen that the way that the sensor works is similar to our approach to get aliasing in previous steps. Indeed, the red and blue images are subsampled by a factor of two in both directions, and the green image is subsampled, but losing only half of the samples.

Therefore, the proposed idea is to reconstruct each color image as color reconstruction algorithms do, obtaining pixel resolution using super-resolution. This idea is presented in **Figure 12**. In this application, each color image is treated as the luminance signal was before, but the motion estimation is done based only on the green values. In this way, we can use SR to replace the color reconstruction algorithm. The first image in the sequence is equivalent to standard reconstruction, but when new images come to the system, the quality increases with the new incoming information.

To avoid a large amount of calculations, the motion vectors are computed only over the green values, and the same motion vectors are applied to the red and blue matrixes. Another reason to do it in that way is that different motion vectors for different colors can produce color aliasing due to different calculated movements of each color. For this reason, the motion vectors must be coherent over the three color planes.

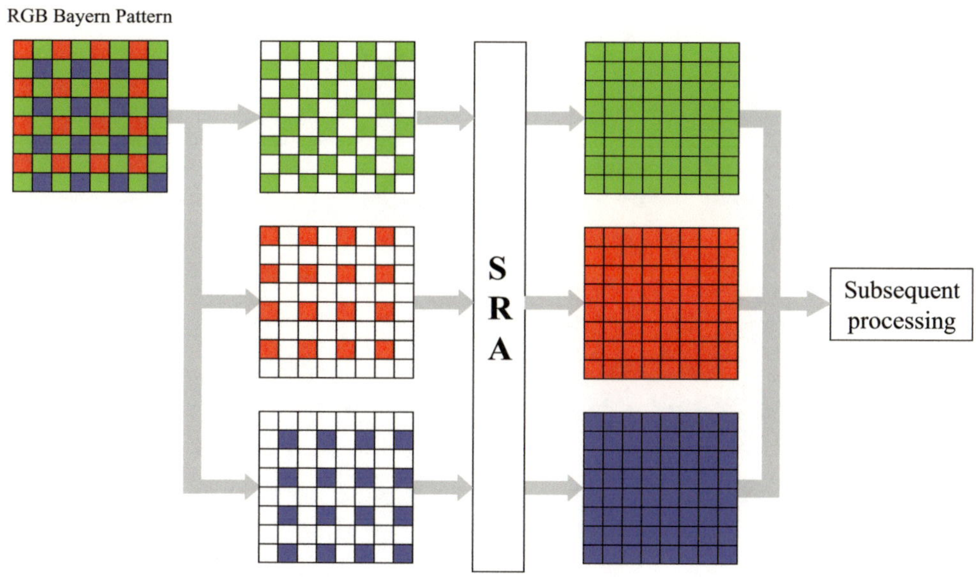

Figure 12. Color reconstruction with super-resolution with aliasing (SRA).

In this case, SR achieves good results. The images obtained are comparable with SmartGreen3 ones, but there are some issues that remain to be tackled:

- In contrast with traditional color reconstruction algorithms, more than one image is needed to increase the quality.
- Errors in the motion estimation will produce bad results. The motion estimation algorithm must be very accurate, or it has to be used only in certain scenes.
- A large number of calculations are needed to perform the motion estimation.

On the other hand, the most important advantages obtained are:

- It works better in colorful images, while SmartGreen use to shift the false colors to the gray scale.
- In case of good motion estimation, the color aliasing is better suppressed.

Moreover, another approach was tested, where the super-resolution with aliasing (SRA) is applied after the overall processing in the image. This processing increases the differences between the values in the RGB domain. For example, the matrix correcting process and the gamma correction provide more differentiate values. This can be useful for the motion estimator to increase the accuracy of the motion vectors. In this case, the proposed process is to follow the next steps:

- Reconstruct the color using standard reconstruction algorithms, keeping the original values. This is necessary because other processing need R, G, and B data in all the positions.
- Do all the processing: matrix correction, gamma correction, white balance, etc.

Figure 13. Standard color reconstruction.

Figure 14. SmartGreen3 color reconstruction with false color detection and edge enhancement.

http://dx.doi.org/10.5772/intechopen.71262

Figure 15. Super-resolution color reconstruction.

- Take the sampled values from the original positions. These values are only modified by the previous processing and not from the color reconstruction algorithms. Using these values, we execute the super-resolution algorithm and reconstruct the lost colors.

With this approach, the obtained results are shown in **Figures 13** and **14**. As it is usual in this kind of processing, where no reference image is available, the results are only qualitative, but in **Figure 15**, a better color reconstruction is appreciated, with less color aliasing in all the colors, but in special in the green color, which is less shifted to the gray scales.

7. Color reconstruction and resolution enhancement

Following the line of the last two applications of Sections 5 and 6, it is possible to join both proposals in a single one. The new idea is to increase the pixel resolution, using each color signal instead of the luminance signal.

In Section 3, it was shown that there are two main sources of aliasing: the color sampling and the fill factor of the sensor. Using these concepts, we can exploit both sources of aliasing in a new application. This new application will deal with color reconstruction and zoom using SR, according to the scheme presented in **Figure 16**. Taking into account a single basic block, 16 pixels are reconstructed from two green values and from one blue and red value. Therefore, more images will be necessary to obtain a good quality, eight in the ideal case of having all the missing positions available.

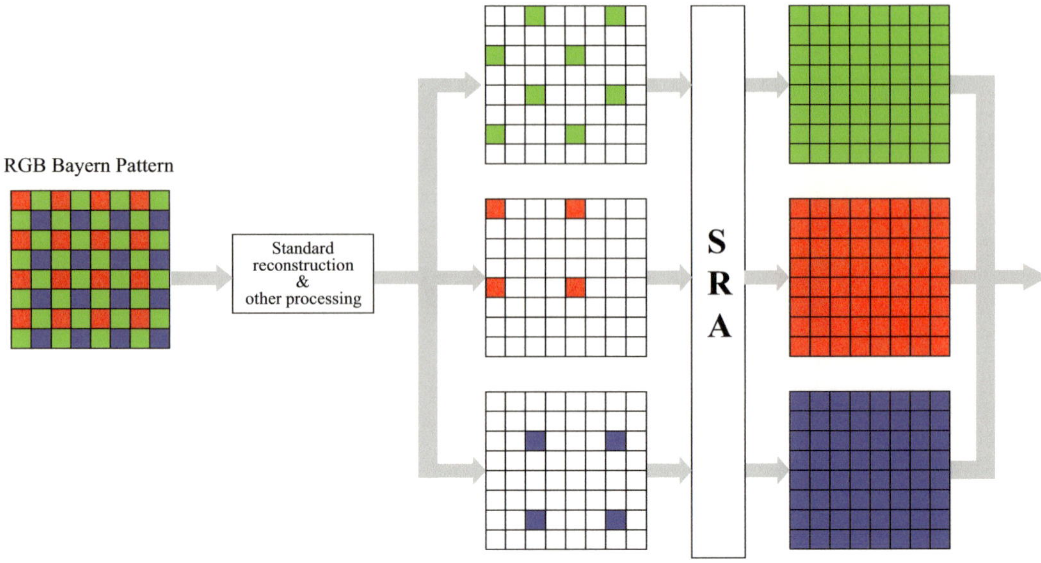

Figure 16. Color reconstruction and zoom using super-resolution.

The processing followed in this case consisted in:

- Reconstructing the color using a standard reconstruction algorithm, keeping the original values.
- Performing all the other processing: matrix correction, gamma correction, white balance, etc.).
- Taking the sampled values from the original positions. These values are only modified by the previous processing and not from the color reconstruction algorithms.
- Applying SR, using a 4 × 4 matrix instead of a 2 × 2 as in the previous cases.

Figure 17. Standard color reconstruction and bilinear interpolation (left). SmartGreen3 color reconstruction followed by false color detection and bilinear interpolation (right).

http://dx.doi.org/10.5772/intechopen.71262

In this case, the motion vectors are also calculated using the green values and reused for the other color matrixes. The usual way to obtain these images is performing first a color reconstruction algorithm followed by a bilinear interpolation. This approach was used in the images of **Figures 17** and **18**. These images are used as a subjective reference to compare against the other images. Following our proposal, some experiments were developed, obtaining the results shown in **Figure 19**.

The obtained results exhibited good quality, although some advantages and disadvantages were found in this new approach. The main disadvantages are:

- More than one image is required to obtain good quality. Nevertheless, this is not a problem for video sequences.
- Bad motion estimation can reduce its quality to bilinear interpolation quality, worse than SmartGreen3 quality.
- It works in RGB data, and it is still necessary to convert to YUV format in some cases, as in video compression.
- It is necessary to suppress the aliasing when super-resolution is not performed.

Also, some advantages are appreciated:

- Better resolution and suppression of color aliasing, achieving a good electronic zooming.
- Reuse of the existing hardware in combination with video compression, for instance, the motion estimation unit.
- Take advantage of undesirable characteristics as aliasing and hand motion.
- Use of signal processing instead of mechanical parts to perform the zoom, contributing in this way to reduce the overall device power dissipation.

Figure 18. SmartGreen3 color reconstruction followed by false color detection, edge enhancement, and bilinear interpolation (left). Standard color reconstruction followed by edge enhancement and bilinear interpolation (right).

Figure 19. Super-resolution reconstruction and zoom (left). Super-resolution reconstruction, zoom, and edge enhancement (right).

8. Conclusions

In this chapter, several applications have been addressed using a real acquisition system. A study of the system has been carried out, analyzing the processing flow in order to identify the possible sources of aliasing. An important problem in this stage is the absence of a reference image to be used in order to obtain quantitative metrics.

Three main applications related with electronic cameras and the image processing and acquisition chain have been addressed:

- *Resolution enhancement*, where three available color reconstruction algorithms have been studied in order to keep the aliasing at the input. As these algorithms do not allow passing a great amount of aliasing, the SR enhancement is not very noticeable.
- *Color reconstruction*, where it is exposed how to apply SR as a way to reconstruct the missing colors when a single sensor is used with a CFA.
- *Color reconstruction and resolution enhancement*, where the two previous applications are combined. In this last case, good results are achieved, making this kind of techniques for such applications recommendable.

Author details

Eduardo Quevedo Gutiérrez[1,2]* and Gustavo Marrero Callicó[1]

*Address all correspondence to: equevedo@iuma.ulpgc.es; eduardo.quevedo@plocan.eu

1 Institute for Applied Microelectronics, University of Las Palmas de Gran Canaria, Las Palmas, Spain

2 Oceanic Platform of the Canary Islands, Carretera de Taliarte, Las Palmas, Spain

References

[1] Tiang J, Ma K-K. A survey on super-resolution imaging. Journal of Signal, Image and Video Processing. 2011;**5**(2):329-342

[2] Goto T, Kawamoto Y, Sakuta Y, Tsutsui A, Sakurai M. Learning-based super-resolution image reconstruction on multi-core processor. IEEE Transactions on Consumer Electronics. 2012;**3**:941-946

[3] Islam MM, Asari VK, Islam MN, Karim MA. Super-resolution enhancement technique for low resolution video. IEEE Transactions on Consumer Electronics. 2010;**56**(2):919-924

[4] Stancil BA, Zhang C, Tsuhan C. Active multi-camera networks: From rendering to surveillance. IEEE Journal of selected topics in signal processing. 2008;**2**(4):597-605

[5] Qinzong P, Kehtarnavaz N. A fast print-from-video technique for cell-phone cameras. IEEE Transactions on Consumer Electronics. 2008;**53**(4):1237-1246

[6] Quevedo E, de la Cruz J, Callicó GM, Tobajas F, Sarmiento R. Video enhancement using spatial and temporal super-resolution from a multi-camera system. IEEE Transactions on Consumer Electronics. 2014;**60**(3):420-428

[7] Quevedo E, de la Cruz J, Sánchez L, Callicó GM, Tobajas F. Super resolution with adaptive macro-block topology applied to a multi camera system. IEEE Transactions on Consumer Electronics. 2015;**61**(2):230-235

[8] Huang TS, Tsay RY. Multiple frame image restoration and registration. In: Huang TS, editor. Advances In Computer Vision and Image Processing. Vol. l. Greenwich, CT: JAI Press Inc.; 1984. p. 317-339

[9] Callicó GM, Peset R, López S, López JF, Núñez A, Sethuraman R, Sarmiento R. Low-cost super-resolution algorithms implementation over a HW/SW video compression platform. EURASIP Journal on Applied Signal Processing. Special Issue on Super-Resolution. 2006;**10**:1-29

[10] Callicó GM, López S, Sosa O, López J, Sarmiento R. Analysis of fast block matching motion estimation algorithms for video super-resolution systems. IEEE Transaction on Consumer Electronics. 2008;**54**(3):1430-1438

[11] Quevedo E, Delroy E, Callico GM, Tobajas F, Sarmiento R. Underwater video enhancement using multi-camera super-resolution. Optics Communications. Elsevier. 2017;**404**: 94-102

[12] Barreto D, Álvarez LD, Molina R, Katsaggelos A, Callicó GM. Region-based super-resolution for compression. Multidimensional systems and signal processing. Special issue on papers presented at the I International Conference in Super Resolution. 2007;**18**:59-81

[13] Parker JA, Kenyon RV, Troxel DE. Comparison of interpolation methods for image resampling. IEEE Transactions on Medical Imaging. 1983;**2**:31-39

[14] Jain AK. Fundamentals in Digital Image Processing. Prentice-Hall; Upper Saddle River, NJ, USA, 1989

[15] Pratt WK. Digital Image Processing. Wiley-Interscience Publishing; Bristol, SOM, United Kingdom, 1991

[16] Zhou J, Zhang H. Super-resolution reconstruction of color image based on microarray lens. International Conference on Applied System Innovation. 2017;**1**:830-833

[17] Antonie C, Jaspers M. Green reconstruction for image sensors. Patent US 7081919 B2, 2006

Chapter 4

Color Analysis and Image Processing Applied in Agriculture

Jesús Raúl Martínez Sandoval,
Miguel Enrique Martínez Rosas,
Ernesto Martínez Sandoval,
Manuel Moisés Miranda Velasco and
Humberto Cervantes De Ávila

Additional information is available at the end of the chapter

http://dx.doi.org/10.5772/intechopen.71539

Abstract

Color and appearance are perhaps the first attributes that attract us to a fruit or vegetable. Since the appearance of the product generally determines whether a product is accepted or rejected, measuring the color characteristics becomes an important task. To carry out the analysis of this key attribute for agriculture, it is recommended to use an artificial vision system to capture the images of the samples and then to process them by applying colorimetric routines to extract color parameters in an efficient and nondestructive manner, which makes it a suitable tool for a wide range of applications. The purpose of this chapter is to give an overview on recent development of image processing applied to color analysis from horticultural products, more specifically the practical usage of color image analysis in agriculture. As an example, quantitative values of color are extracted from Habanero Chili Peppers using image processing; the images from the samples were obtained using a desktop configuration of machine vision system. The material presented should be useful for students starting on the field, as well as for researchers looking for state-of-the-art studies and practical applications.

Keywords: color analysis, color evolution, feature extraction, image processing, Habanero chili, computer vision

1. Introduction

The color assessment of fruits and vegetables in the food industry and agriculture using machine vision and image processing has become a trend in the recent years [1–3]. The color features are one of the key parameters to define the quality of an agricultural product [4, 5]. The color is probably the first factor that consumers use to determine the appearance of a product [6]; appearance is a subjective factor that leads the consumer to accept or reject a food product [7]. This significantly affects the sales and profits of the industry. Therefore, a considerable effort has been made in the area of automation to improve the quality of agricultural products in the food industry in order to decrease losses.

Building machines with the ability to see color as it does the human being has been a complex task for the scientific community and industry in recent years [8–12]. Among the many challenges to be addressed, we can include appropriate image acquisition systems, lighting problems, color space definitions, mathematical issues, and the development of specific algorithms and synchronization tasks [13]. However, improvements in semiconductors, electronics, and software eventually brought the opportunity to implement image processing and colorimetric projects for various applications [11, 14, 15].

A machine vision system for horticulture products requires the ability to capture process and analyze color images, where algorithms are suitable to detect, extract, and quantify the attribute of color as much as a customer does. Furthermore, other parameters (size, texture, external blemishes, and diseases) are important to determine the appearance, and hence the quality, of an agricultural product [16–18]. This variety of applications is possible due to the interaction of light in the range of visible spectrum (400–700 nm) with the matter that the light can be reflected, transmitted, or absorbed by an object. The light wavelengths received by our eyes are then interpreted by our brain as color.

Perception of color, in humans, is a psychophysical phenomenon that involves three elements: the illuminant, an object, and the observer [19]. The illuminant has the function to irradiate the object with light in the visible spectrum. The object absorbs, transmits, and reflects the light received from the illuminant. The observer perceives the reflected light from the object in the retina and responds to that stimulus generating an electric signal in the optic nerve toward the brain [20]. As this phenomenon takes place in the brain, the process is an important challenge, because machine vision systems need to emulate these three elements, in a proper way, without including the real observer, the eye, and the natural illuminant, the sun [21].

A machine vision system acquires, processes, and analyzes images, and the proper operation of the system requires adequate capture conditions for the specific application. Basic machine vision components are: (a) image sensor (charge-couple device (CCD) or complementary metal oxide semiconductor (CMOS)), (b) illuminant (D-65, fluorescent lamps, among others), (c) the background (with high contrast for the object of interest), (d) trigger device (used to start image acquisition), and (e) a frame grabber (to capture the actual image). The capture procedure can be divided into the following steps: the illuminator irradiates the object, the image sensor receives the light reflected from the object and the background, and when the trigger is activated, the system extracts the color characteristics and converts them to electrical signals and then the grabber frame stores the image taken [22, 23]. Once the image is on the computer or other processing device, image processing algorithms are used to analyze the data. The basic steps can be listed as: (a) segmentation of images (the background is separated from the region of interest), (b) extraction of characteristics (select pixels of interest); for this stage, several functions and algorithms have been developed in order to get an adequate image processing of the scene. Next, we analyze the data generated from the previous process, the system extracts the color attribute of the objects in the image using colorimetric techniques, as well as others as morphological and texture parameters [24–26], and finally the analysis results are presented.

Colorimetry is the scientific area that measures, quantifies, and represents color [27]. It is very useful in different areas because it provides the ability to turn color into an objective factor rather than a subjective one. Complementing this scientific area with the technology of computer vision systems, it is possible to see the color characteristics in a digital and standard way, in addition to performing the color evaluation with a noninvasive and non-contact procedure, which makes it a suitable technology for application in agriculture and the food industry for quality assessment. Due to their versatility, many other production systems can benefit from the features offered by artificial vision systems, image processing, and colorimetry.

The aim of this study is to evaluate noninvasive and noncontact techniques of image processing and color analysis; for example, the estimation of color from postharvest Habanero chili at different maturity stages is shown. Moreover, the capability of color assessment as a region, instead of a punctual area as is done with a typical colorimeter, identifies threshold values in these stages.

The goal is to identify and quantify color attributes of Habanero chili fruits using the following stages:

- Image acquisition system (setup).
- Image processing (threshold technique and algorithms).
- Feature extraction (pixels of interest and color calculations).
- Color analysis (attribute identification, quantification and classification).

2. Image processing

The techniques of image processing are related to algorithms that manipulate the numerical representation of the images to obtain useful information [28]. In particular, segmentation subdivides the digital image into multiple regions or objects that have common characteristics [29]. This action involves several steps that must be taken, before the images can provide valuable data. First, region of interest (ROI) must be identified, i.e., regions in the image that have pixels that matter to the application and which must be separated from the background. In order to develop algorithms that perform this work, it is necessary to define whether the image is black and white or colored, and then a criterion must be established for the segmentation, which must focus on morphological, texture, or colorimetric parameters, before establishing any flow chart as solution. It should be mentioned that algorithms may be suitable for many applications, but none of them is generally applicable to all images, and therefore, suitable algorithms for each particular application should be used [30, 31].

2.1. Image acquisition system and color image representation

It is essential to use an image acquisition system that suits the application properly. It must be defined what type of objects or products will be placed in the system, as well as set the scene location and choose the appropriate settings to obtain acceptable images. The issue of lighting (natural or artificial) depends on the scenes required by the application. Once the capture conditions are adequate, a vision sensor (CCD or CMOS) must be selected, which will suit the entire system for proper performance. A digital camera receives light variations corresponding to images onto a CCD device. The CCD contains capacitors that are stimulated by visible radiation and three filters adjusted for three basic colors: red (R), green (G), and blue (B). Theoretically, every color can be reproduced by the combination of three primary colors.

A color image are represented as a M × N × 3 (components) array of color pixels, where each color pixel is a triplet corresponding to the RGB components of and image at specific spatial location, as shown in **Figure 1**. By convention, the three images formed and RGB color image are referred to as the R, G, B component images. The data class of the component from the image determines their ranges of values, for example [0–255] or [0–65,535] for RGB images of class uint8 or uint16, respectively [32]. The RGB color space is an additive color model that uses transmitted light to display colors. It is used for television and other devices screens; so this model is device dependent (Its appearance depends on the display.)

A summary of machine vision systems, image processing, and color analysis frequently implemented in the agriculture is presented in **Table 1**. First column corresponds to the type of application, and four general and typical areas are included. The type of tasks performed by these applications is shown in the second column, including the corresponding references. Third column shows the type of setup employed for each application, and the last column shows the locations where these systems are typically deployed.

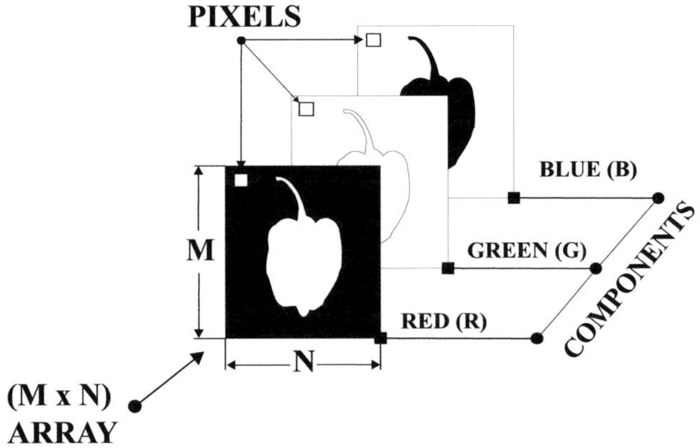

Figure 1. Color representation of digital image and their RGB components.

Application	Task	Type of setup	Location
In situ	Monitoring Color Diseases Plague Growth Behavior [33–36]	Portail By satellite Aerial inspection [37–42]	On field
Harvesting robots	Harvest Shape Color [43–46]	Mobile robot	On field
Online sorting	Sorting Color Size Quality Blemish [47–49]	Static machine (placed on a production line)	Packer station
Postharvest	Analysis Quality Shelf life Volume Blemish [8, 50–54]	Desktop	Laboratory

Table 1. Image acquisition setups and their typical usage.

2.2. Color image segmentation

Algorithms for color image segmentation based on the technique of threshold segmentation should be executed in triplicate, due to the structure of the color image [55, 56]. A typical segmentation process fills in with ones and zeros of the image matrix locations, corresponding to the selected regions in each of the color channels of an image, as shown in **Figure 2**.

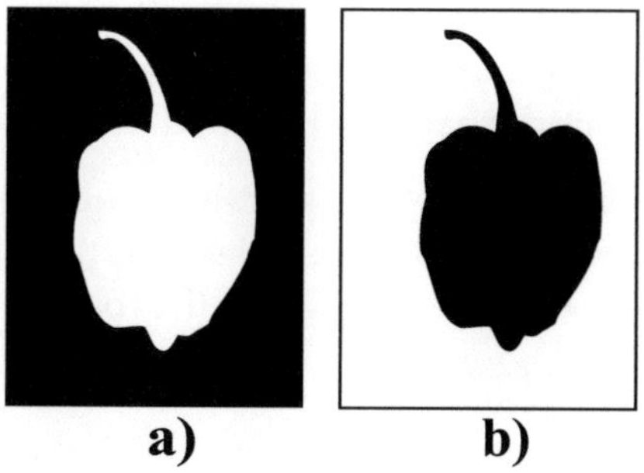

Figure 2. Segmented image: (a) background filled with zeros and (b) background filled with ones.

3. Color calculations

The Commission Internationale de l'Eclairage (CIE) determines regulations, standards, and recommendations for color measurements. The CIELAB color space is an international standard developed by the CIE in 1976. Within CIELAB, a psychometric index of lightness (L^*) and two color coordinates (a^* and b^*) are defined. L^* is a qualitative attribute of relative luminosity, which is the property according to which each color can be considered as equivalent to a member of the gray scale, ranging between black ($L^* = 0$) and white ($L^* = 100$). Negative values from a^* correspond to greenish and the positives to the reddish ones, whereas the yellowish colors takes negatives values from b^* and the positive for bluish ones [57].

It is well known in the food industry that the CIELAB color space is used to analyze color changes in a qualitative way [58–62]. Color and appearance are closely related to the sensory properties and chemical composition of food. Color is usually measured by tristimulus colorimetry. The color stimulus is composed of three different parameters, giving the color a three-dimensional nature.

The color attributes are described in CIELAB as:

- Lightness: This feature indicates if a color is lighter or darker. It is a relative measure of the reflected light against the absorbed. Value 0 corresponds to black and value 100 is assigned to white.
- Chroma: It determines for each hue, the color difference taking as reference the gray level with same lightness. It can take positive values from zero.
- Hue: It is the main attribute. It is a qualitative property, which allows classifying colors as red, yellow, etc. It is related to differences in absorbance of radiant energy at different wavelengths. Hue is specified as an angle.

These attributes are often expressed as L^*, C_{ab}^*, and h_{ab}°, respectively; according to the CIELAB color space, it can be represented as Cartesian coordinates of polar coordinates C_{ab}^* and h_{ab}°. It can be used on a variety of instruments, such as colorimeters, spectrophotometers, and spectroradiometers. However, these instruments require homogenous samples to achieve a uniform color, which becomes a tedious and complicated task to measure the color of heterogeneous or small objects, such as grape berries and grape seeds. In these cases, the use of digital images for the extraction of color characteristics is advantageous. Digital image analysis appears as a suitable complement since it is possible to extract not only color characteristics but also other characteristics such as shape, texture, and homogeneity.

On the other hand, due to the nature of the CIELAB color space, the calculation of the Euclidean distance can be applied between neighboring samples, in order to obtain a relation between a quantitative and qualitative value on the variation of the color appearance of an object.

4. Application of image processing and color analysis of Habanero chili pepper

Since consumers buy with their eyes, color is considered one of the most important quality parameters of food products. Normally, this is determined by human inspection, or measured using a colorimeter or a spectrophotometer. The first process is subjective and susceptible to fatigue. The second is limited to measure just a small area of the food product, making it difficult to obtain a clear view of the color of the complete sample [63]. In order to overcome these limitations, a system of artificial vision, image processing, and color analysis has been applied to measure the postharvest color of the Habanero chili (*Capsicum chinense* Jacq.) at different stages of ripening.

4.1. Genus capsicum

The Habanero chili belongs to the family Solanaceae and genus Capsicum. This genus consists of 27 species, five of which have been domesticated and are used worldwide as vegetables, spices, and condiments: *Capsicum annum L., Capsicum frutescens L., Capsicum chinense J., Capsicum pubescens R. & P.*, and *Capsicum baccatum L.*, where nonpungent cultivars of *C. annum* are the most consumed and are the main objective of most breeding programs [64].

Chili peppers (Capsicum spp.) are well known for their ability to cause an intense organoleptic sensation of heat when consumed (pungency). Capsaicin and its analogues, collectively called capsaicinoids (a group of alkaloids), are responsible for giving pungency or heat to the fruit; the pungent feature of peppers is only present in the members of the genus Capsicum [65–67].

The *Capsicum chinense* Jacq. is a very aromatic pepper and also is one of the hottest peppers in the world [68]. These fruits are commonly used to give a pungent or hot sensation to many different meals and food products all around the world. During the past decade, it has been reported that the consumption of certain foods and spices such as pepper may have a positive effect on health. Genus Capsicum shows an incredible diversity and is consumed by a large section of population throughout the world because of its impressive health beneficial chemical compounds such as capsaicinoids, carotenoids (provitamin A), flavonoids, vitamins (vitamins C and E), minerals, essential oils, and aroma of the fruits. These compounds have been shown to possess anticancer, anti-inflammatory, antimicrobial, and antioxidant properties [69]. They are important from the economic point of view for countries like Mexico, not only for the preparation of regional foods that have made Mexican gastronomy famous all over the world, but also because of the great amount and genetic variability that occurs in its territory, especially spicy species [68, 70–73].

4.2. Methodology

The methodology used is described as follows: (1) specimens of Habanero chili here collected and arranged in three groups depending on their maturity stage by visual inspection. (2) The samples are moved to the laboratory to be placed in the machine vision system and the acquisition process is setup. (3) The acquisition process starts with a 24 sampling rates and finishes after 15 days, capturing a single image from each specimen. (4) Once the images were acquired, a database with a total of 900 images was generated. (5) The algorithms of image processing and color analysis were applied to this dataset. 6) Finally, the algorithms generate results with color segmentation, colorimetric measurements in CIELAB color space, color analysis, and statistical analysis of the dataset.

4.2.1. Samples

For this case of study samples of Habanero chili were harvested from an aquaponic greenhouse culture. The selection of the samples was carried out by a specialized technician. The expert harvested a representative group of Habanero chili, which showed different stages of maturity. A color categorization was performed by visual inspection. The samples were separated into three groups of colors (green, yellowish and orange), with 20 specimens per group.

4.2.2. Setup of machine vision system

The artificial vision system used a color CCD camera as image capture device. The lighting system contained fluorescent lamps mounted on top to avoid shadows. In order to control the camera and to download the images, we used a standard PC with MATLAB. The system design allowed the camera to move the samples above to keep them stable. Because of the protrusions and cavities presented by the Habanero chili, it is difficult to keep the surface of interest in exactly the same position along the experiment. As shown in **Figure 3**, the samples were placed in the confined space of the machine vision system to acquire the corresponding images.

4.2.3. Image processing algorithms

To accomplish the image processing tasks, the flow chart was followed as showed in **Figure 4**. The algorithms are executed using suitable functions in order to get reliable information from digital images. At the beginning, the image from sample is captured using the image acquisition system. Then, image processing routines carry out the required operations to separate the ROI, as well as to convert the image from RGB color space to CIELAB. The calculations to obtain the CIELAB components (L*, a*, b*, Chroma, Hue angle) using tristimulus colorimetry are performed to generate the corresponding array. Finally, the color analysis from the Habanero chili in postharvest conditions can be accomplished.

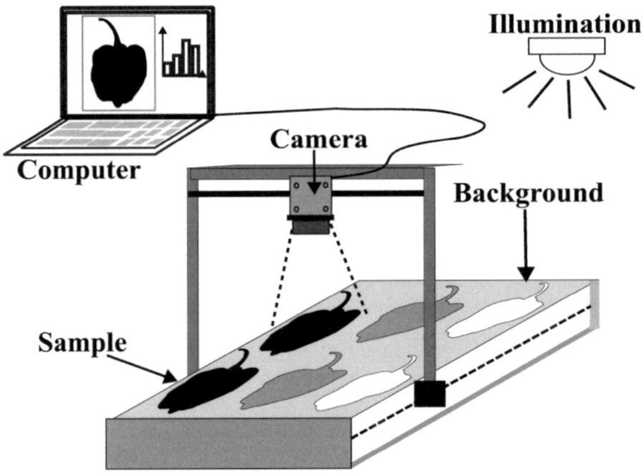

Figure 3. Proposed setup for the case of study.

http://dx.doi.org/10.5772/intechopen.71539

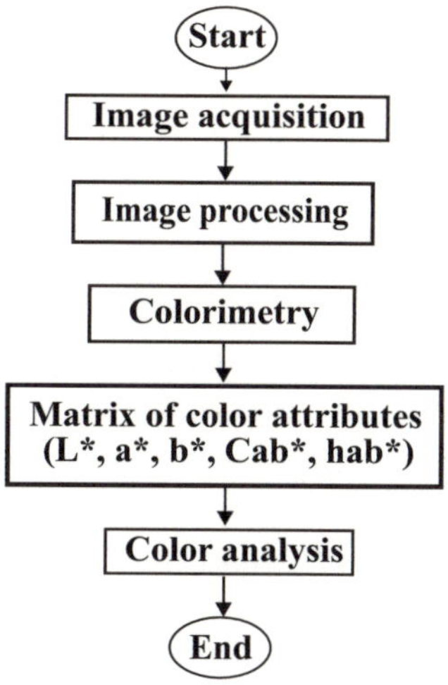

Figure 4. Flow chart used for the image processing algorithm.

The original and processed images, by the algorithms from the dataset, are shown in **Figure 5**. One image, aleatory selected, from each categorized group (green, yellowish, and orange) is presented. At the top, you can identify the categorization of groups, and below is the corresponding label of the sample. Then, the original images, just as the vision system acquired them, are presented. In the middle, the label that indicates the exact day of the acquisition is located. At the bottom, the processed images are shown, where you can clearly observe the segmentation process and how the algorithms filled the background with zeros, and only the information from the region of interest is processed (ROI).

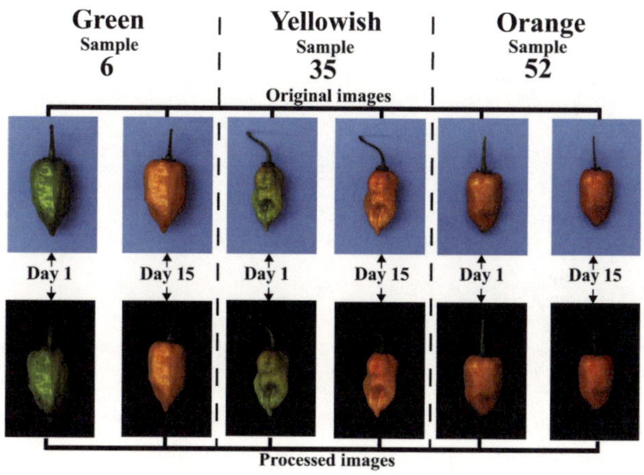

Figure 5. Original and processed images of Habanero chili.

4.2.4. Color and statistical analysis

Habanero chili is a climacteric fruit, which means that once it is cut, it begins to ripen. Depending on the variety of *Capsicum chinense* Jacq., the color of the Habanero chili changes during maturation. In general, changes occur from the dark green, in its initial stages, and then go through the yellowish-green until reaching an orange color, in the final stage of maturity.

Color evolution can be represented as hue angles (h_{ab}°) in a polar graph. Typically, Habanero chili initiates maturity with green stage (under the threshold of h_{ab} = 120° for green) and then it moves through the yellowish stage (between the threshold h_{ab} = 120 and 60° for yellowish) to achieve orange colors (crossing the threshold of h_{ab} = 60° for orange) in the final stage of maturity, as time passes. The color information from the dataset showed this behavior, and the statistical analysis presented in **Figure 6** demonstrates that image processing and colorimetry are capable of extracting reliable values from acquired images and detect color changes from these agriculture products. Therefore, this methodology described, as noncontact technique, can be considered a suitable option to analyze the color of Habanero chili.

In **Figure 6**, chart (a) shows the variation from the green group. In average values of hue angle, a high color change is presented, due to the gradual transition from green colors to oranges passing through the yellowish ones during their maturity process. In chart (b), a descendant gradual color change can be appreciated, even when crossing the threshold of

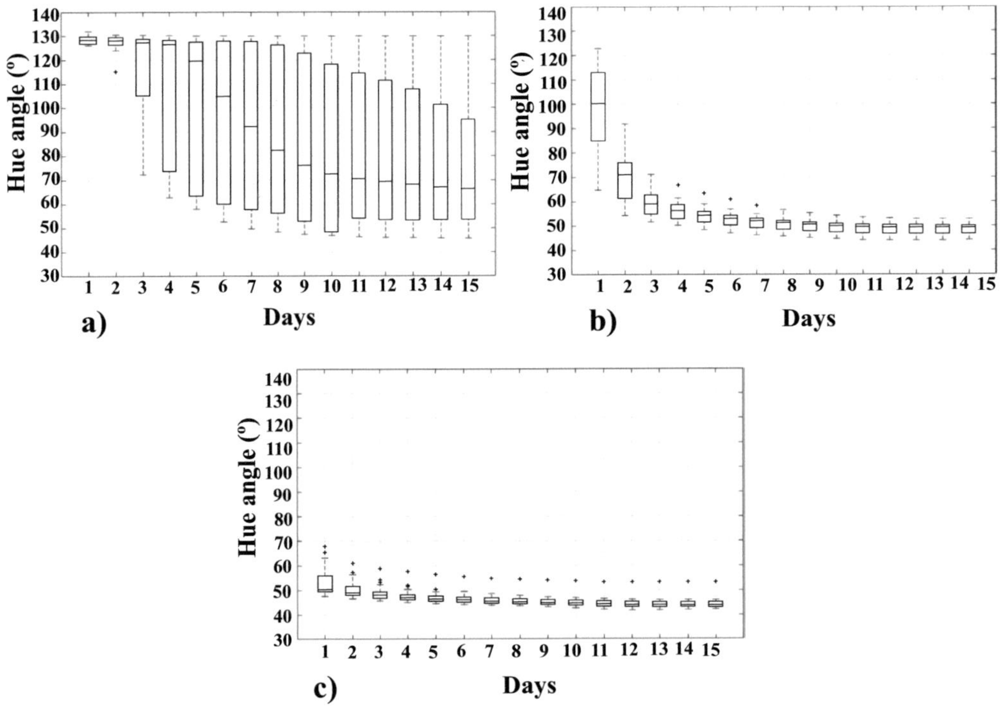

Figure 6. Box and whiskers chart from each group: (a) green, (b) yellowish, and (c) orange.

Source	SS	DF	MS	$F > F_{0.05}$
Green				
Days	91722.7	14	6551.62	8.17 >> 1.692
Error	228586.7	285	802.06	—
Total	320309.4	299	—	—
Yellowish				
Days	49264.8	14	3518.91	100.21 >> 1.692
Error	10008.2	285	35.12	—
Total	59,273	299	—	—
Orange				
Days	1915.77	14	136.841	14.6 >> 1.692
Error	2670.3	285	9.369	—
Total	4586.07	299	—	—

Table 2. One-way ANOVA table for each group.

h_{ab} = 60°. Instead, the third group (c), corresponding to the orange ones, remains after the threshold of h_{ab} = 60° with a slow progressive color changes.

A one-way ANOVA was conducted to evaluate the relationship between color changes in 20 Habanero chili and 15 days of sampling. **Table 2** displays the summary for the one-way ANOVA for each group of samples. With the null hypothesis (H_0), it was showed that all the color values are equal during 15 days. In the analysis, it was shown that in green, yellowish, and the orange groups, the ANOVA was significant: F(14, 285) = 8.17, F(0.05) = 1.692, F(14, 285) = 100.21, F(0.05) = 1.692, and F(14, 285) =8.17, F(0.05) = 1.692, respectively. The ANOVA results allowed to reject the null hypothesis and supported the conclusion that there is a statistically significant color change during days for the green, yellowish, and orange groups.

5. Conclusions

Artificial vision systems combined with image processing and color analysis are a reliable and affordable option when specific applications require the use of noninvasive and noncontact techniques. Similar characteristics of the samples are extracted from their images and grouped for further analysis using image processing techniques, which helps to obtain consistent and reliable separations of elements. The CIELAB color space provides the parameters needed to analyze and calculate important characteristics of a color image. Color differences can be detected more directly using the CIELAB color space, and it is important to mention that a color difference magnitude can be imperceptible to the naked eye, but it is a basic operation for the vision systems. However, the context in the color analysis should be considered as an important factor, for the proper interpretation of the data generated from the previous process. For example, the color attribute of the Habanero chili is a fundamental parameter for the appearance of the genus capsicum, which can be evaluated by image processing and colorimetry to detect color changes with adequate and reliable results in postharvest analysis. Trend in applications of color analysis and image processing for agriculture will continue to increase in the near future, due to the great variety of colors and shapes of the products, in particular, the interest to obtain the best quality.

Acknowledgements

The authors wish to thank the National Council for Science and Technology (CONACyT) for providing the funding to carry out this research, as well as to the people involved in this research at the State University of Sonora, the Autonomous University of Baja California, Program for Strengthening Educational Quality (PFCE) and the Center for Scientific Research and Higher Education at Ensenada.

Author details

Jesús Raúl Martínez Sandoval[1]*, Miguel Enrique Martínez Rosas[2], Ernesto Martínez Sandoval[2], Manuel Moisés Miranda Velasco[2] and Humberto Cervantes De Ávila[2]

*Address all correspondence to: jesus.martinez@ues.mx

1 State University of Sonora, Hermosillo, Mexico

2 Autonomous University of Baja California, Ensenada, Mexico

References

[1] Zhang B, Huang W, Li J, Zhao C, Fan S, Wu J, Liu C. Principles, developments and applications of computer vision for external quality inspection of fruits and vegetables: A review. Food Research International. 2014. DOI: 10.1016/j.foodres.2014.03.012

[2] Ayman HAE, Ayman AAK. Understanding color image processing by machine vision for biological materials. In: Structure and function of food engineering. INTEHC; 2012. pp. 227-274. DOI: 10.5772/50796

[3] Feng Y-Z, Sun D-W. Application of Hyperspectral imaging in food safety inspection and control: A review. Critical Reviews in Food Science and Nutrition. 2012;**52**(11):1039-1058. DOI: 10.1080/10408398.2011.651542

[4] Saldaña E, Siche R, Luján M, Quevedo R. Review: Computer vision applied to the inspection and quality control of fruits and vegetables. Revisão: visão computacional aplicada à inspeção e ao controle da qualidade de frutas e verduras. Brazilian Journal of Food Technology, Campinas. 2013;**16**(4):254-272. DOI: 10.1590/S1981-67232013005000031

[5] Pathare PB, Opara UL, Al-Said FAJ. Colour measurement and analysis in fresh and processed foods: A review. Food and Bioprocess Technology. 2013;**6**(1):36-60. DOI: 10.1007/s11947-012-0867-9

[6] Maldonado A, Vigneault C, Nakaji K. Postharvest Technologies of Fresh Horticulture Produce. In: Horticulture, 2012, pp. 164-172, InTech, ISBN: 9789535102526

[7] G Hareesh K S SL,NV. Quality inspection and grading of agricultural and food products by computer vision-a review. International Journal of Computer Applications. 2010;**2**(1):975-8887. DOI: 10.5120/612-863

[8] Wabali VC, Esiri A, Zitte L. A sensory assessment of color and textural quality of refrigerated tomatoes preserved with different concentrations of potassium permanganate. Food Science and Nutrition. 2017;**5**(3):434-438. DOI: 10.1002/fsn3.410

[9] Ji B, Wang J, Liu W. Color-based automatic quality control for roasting chicken. Computers and Electronics in Agriculture. 2016;**123**:49-56. DOI: 10.1016/j.compag.2016.01.028

[10] Singh H, Gill N. Machine vision based color grading of Kinnow mandarin. International Journal of Advanced Research in Computer Science and Software Engineering. 2015;**5**(5):1253-1259

[11] Schmilovitch Z, Ignat T, Alchanatis V, Gatker J, Ostrovsky V, Felföldi J. Hyperspectral imaging of intact bell peppers. Biosystems Engineering. 2014;**117**(C):83-93. DOI: 10.1016/j.biosystemseng.2013.07.003

[12] Krig S. Computer vision metrics: Survey, taxonomy, and analysis. Computer Vision Metrics: Survey, Taxonomy, and Analysis. 2014;**9781430259**:1-472. DOI: 10.1007/978-1-4302-5930-5

[13] Morene OC., Gila DMM, Puerto DA, García JG, Ortega JG. Automatic determination of peroxides and acidity of olive oil using machine vision in olive fruits before milling process. IST 2015-2015 IEEE International Conference on Imaging Systems and Techniques, Proceedings; 2015. DOI: 10.1109/IST.2015.7294543

[14] Rahman MO, Hussain A, Scavino E, Basri H, Hannan M a. Intelligent computer vision system for segregating recyclable waste papers. Expert Systems with Applications. 2011;**38**(8):10398-10407. DOI: 10.1016/j.eswa.2011.02.112

[15] Kang, S. P., East AR, Trujillo FJ. Colour vision system evaluation of bicolour fruit: A case study with "B74" mango. Postharvest Biology and Technology. 2008;**49**(1):77-85. DOI: 10.1016/j.postharvbio.2007.12.011

[16] Barrett DM, Beaulieu JC, Shewfelt R. Color, flavor, texture, and nutritional quality of fresh-cut fruits and vegetables: Desirable levels, instrumental and sensory measurement, and the effects of processing. Critical Reviews in Food Science and Nutrition. 2010;**50**(5):369-389. DOI: 10.1080/10408391003626322

[17] Chauhan APS, Singh AP. Intelligent estimator for assessing apple fruit quality. International Journal of Computer Applications. 2012;**60**(5):35-41

[18] Zareiforoush H, Minaei S, Alizadeh MR, Banakar A. Qualitative classification of milled rice grains using computer vision and metaheuristic techniques. Journal of Food Science and Technology. 2016;**53**(1):118-131. DOI: 10.1007/s13197-015-1947-4

[19] González IA, Osorio C, Meléndez-Martínez AJ, González-Miret ML, Heredia FJ. Application of tristimulus colorimetry to evaluate colour changes during the ripening of Colombian guava (Psidium Guajava L.) varieties with different carotenoid pattern. International Journal of Food Science & Technology. 2011;**46**(4):840-848. DOI: 10.1111/j.1365-2621.2011.02569.x

[20] Heredia FJ, González-Miret ML, Meléndez-Martínez AJ, Vicario IM. Instrumental assessment of the sensory quality of juices. In: Instrumental assessment of food sensory quality. 2013. pp. 565-610, Woodhead Publishing, ISBN: 9780857094391

[21] Meléndez-Martínez AJ, Gómez-Robledo L, Melgosa M, Vicario IM, Heredia FJ (2011). Color of orange juices in relation to their carotenoid contents as assessed from different spectroscopic data. Journal of Food Composition and Analysis, **24**(6):837-844. DOI: 10.1016/j.jfca.2011.05.001

[22] Rodríguez-Pulido FJ, Gordillo B, Lourdes González-Miret M, Heredia FJ. Analysis of food appearance properties by computer vision applying ellipsoids to colour data. Computers and Electronics in Agriculture. 2013;**99**:108-115. DOI: 10.1016/j.compag.2013.08.027

[23] Rodríguez-Pulido FJ, Gómez-Robledo L, Melgosa M, Gordillo B, González-Miret ML, Heredia FJ. Ripeness estimation of grape berries and seeds by image analysis. Computers and Electronics in Agriculture. 2012;**82**:128-133. DOI: 10.1016/j.compag.2012.01.004

[24] Lu R, Cen H. Non-destructive methods for food texture assessment. In: Instrumental assessment of food sensory quality; 2013. p. 230-255e. DOI: 10.1533/9780857098856.2.230

[25] Lu R. Principles of solid food texture analysis. In: Instrumental assessment of food sensory quality. 2013. p. 103-128. DOI: 10.1533/9780857098856.1.103

[26] Kim DG, Burks TF, Qin J, Bulanon DM. Classification of grapefruit peel diseases using color texture feature analysis. Int J Agric & Biol Eng Open Access at Http://www.ijabe.org. International journal of agricultural and biological engineering. 2009;**2**(23):41-50. DOI: 10.3965/j.issn.1934-6344.2009.03.041-050

[27] Collewet C, Marchand E. Colorimetry-based visual servoing. 2009 IEEE/RSJ International Conference on Intelligent Robots and Systems. IEEE; 2009. DOI: 10.1109/IROS.2009.5354416

[28] Wang XY, Sun WW, Wu ZF, Yang HY, Wang QY. Color image segmentation using PDTDFB domain hidden Markov tree model. Applied Soft Computing Journal. 2015;**29**:138-152. DOI: 10.1016/j.asoc.2014.12.023

[29] Bova N, Ibanez O, Cordon O. Image Segmentation Using Extended Topological Active Nets Optimized by Scatter Search. IEEE Computational Intelligence Magazine. 2013;**8**(1):16-32. DOI: 10.1109/MCI.2012.2228587

[30] Peng B, Li T. A probabilistic measure for quantitative evaluation of image segmentation. IEEE Signal Processing Letters. 2013;**20**(7):689-692. DOI: 10.1109/LSP.2013.2262938

[31] Russ JC. The Image Processing Handbook. Image Processing (Sixth Edit): CRC Press; 2011. DOI: 10.1201/9781420040760

[32] Gonzalez RC, Woods RRE, Eddins SL. Digital Image Processing Using Matlab. Gatesmark Publishing: Gatesmark Publishing; 2004

[33] Mehta SS, Ton C, Asundi S, Burks TF. Multiple camera fruit localization using a parti`cle filter. Computers and Electronics in Agriculture. 2017;**142**:139-154. DOI: 10.1016/j.compag.2017.08.007

[34] Pires RDL, Gonçalves DN, Oruê JPM, Kanashiro WES, Rodrigues JF, Machado BB, Gonçalves WN. Local descriptors for soybean disease recognition. Computers and Electronics in Agriculture. 2016;**125**:48-55. DOI: 10.1016/j.compag.2016.04.032

[35] Li Y, Cao Z, Lu H, Xiao Y, Zhu Y, Cremers AB. In-field cotton detection via region-based semantic image segmentation. Computers and Electronics in Agriculture. 2016;**127**:475-486. DOI: 10.1016/j.compag.2016.07.006

[36] Çakir Y, Kirci M, Güneş EO, Üstündağ, BB. Detection of oranges in outdoor conditions. In: 2013 2nd International Conference on Agro-Geoinformatics: Information for Sustainable Agriculture, Agro-Geoinformatics 2013. 2013. p. 500-503. DOI: 10.1109/Argo-Geoinformatics.2013.6621971

[37] ZHOU, Qing bo, YU, Qiang yi, LIU, Jia, WU, Wen bin, TANG, Hua jun. Perspective of Chinese GF-1 high-resolution satellite data in agricultural remote sensing monitoring. Journal of Integrative Agriculture. 2017;**16**(2):242-251, DOI: 10.1016/S2095-3119(16)61479-X.

[38] Pasher J, McGovern M, Putinski V. Measuring and monitoring linear woody features in agricultural landscapes through earth observation data as an indicator of habitat availability. International Journal of Applied Earth Observation and Geoinformation. 2016;**44**:113-123. DOI: 10.1016/j.jag.2015.07.008.

[39] Liu K, Zhou Q, Wu W, Xia T, Tang H. Estimating the crop leaf area index using hyperspectral remote sensing. Journal of Integrative Agriculture. 2016;**15**(2):475-491. DOI: 10.1016/S2095-3119(15)61073-5

[40] Elarab M, Ticlavilca AM, Torres-Rua AF, Maslova I, McKee M. Estimating chlorophyll with thermal and broadband multispectral high resolution imagery from an unmanned aerial system using relevance vector machines for precision agriculture. International Journal of Applied Earth Observation and Geoinformation. 2015;**43**:32-42. DOI: 10.1016/j.jag.2015.03.017

[41] Rokhmana CA. The potential of UAV-based remote sensing for supporting precision agriculture in Indonesia. Procedia Environmental Sciences. 2015;**24**:245-253. DOI: 10.1016/j.proenv.2015.03.032

[42] Murillo-Bracamontes EA, Martinez-Rosas ME, Miranda-Velasco MM, Martinez-Reyes HL, Martinez-Sandoval JR, Cervantes-De-Avila H. Implementation of Hough transform for fruit image segmentation. Procedia Engineering. 2012;**35**:230-239. DOI: 10.1016/j.proeng.2012.04.185

[43] Wang C, Tang Y, Zou X, SiTu W, Feng W. A robust fruit image segmentation algorithm against varying illumination for vision system of fruit harvesting robot. Optik. 2017;**131**:626-631. DOI: 10.1016/j.ijleo.2016.11.177

[44] Mehta SS, Burks TF. Multi-camera Fruit Localization in Robotic Harvesting. IFAC-PapersOnLine. 2016;**49**(16):90-95. DOI: 10.1016/j.ifacol.2016.10.017

[45] García-Luna F, Morales-Díaz A. Towards an artificial vision-robotic system for tomato identification. IFAC-PapersOnLine. 2016;**49**(16):365-370. DOI: 10.1016/j.ifacol.2016.10.067

[46] Zujevs A, Osadcuks V, Ahrendt P. Trends in robotic sensor Technologies for Fruit Harvesting: 2010-2015. Procedia Computer Science. 2015;**77**:227-233. DOI: 10.1016/j.procs.2015.12.378

[47] Mo C, Kim G, Kim MS, Lim J, Lee K, Lee WH, Cho BK. On-line fresh-cut lettuce quality measurement system using hyperspectral imaging. Biosystems Engineering. 2017;**156**:38-50. DOI: 10.1016/j.biosystemseng.2017.01.005

[48] Jhawar J. Orange Sorting by Applying Pattern Recognition on Colour Image. Physics Procedia. December 2016;**78**(2015):691-697. DOI: 10.1016/j.procs.2016.02.118

[49] Kim MS. Online screening of fruits and vegetables using hyperspectral line-scan imaging techniques. In: High Throughput Screening for Food Safety Assessment. 2014, pp. 467-490. Woodhead Publishing, ISBN: 9780857098078.

[50] Hashimoto A, Muramatsu T, Suehara K, Kameoka S, Kameoka T. Color evaluation of images acquired using open platform camera and mini-spectrometer under natural lighting conditions. Food Packaging and Shelf Life. 2017. DOI: 10.1016/j.fpsl.2017.08.008

[51] Cho JS, Lee HJ, Park JH, Sung JH, Choi JY, Moon KD. Image analysis to evaluate the browning degree of banana (Musa spp.) peel. Food Chemistry. 2016;**194**:1028-1033. DOI: 10.1016/j.foodchem.2015.08.103

[52] Bac CW, Hemming J, Van Henten EJ. Pixel classification and post-processing of plant parts using multi-spectral images of sweet-pepper. IFAC Proceedings Volumes. 2013;**46**(4):150-155. DOI: 10.3182/20130327-3-JP-3017.00035.

[53] Ataş M, Yardimci Y, Temizel A. A new approach to aflatoxin detection in chili pepper by machine vision. Computers and Electronics in Agriculture. 2012;**87**:129-141. DOI: 10.1016/j.compag.2012.06.001

[54] Omid M, Khojastehnazhand M, Tabatabaeefar A. Estimating volume and mass of citrus fruits by image processing technique. Journal of Food Engineering. 2010;**100**(2):315-321. DOI: 10.1016/j.jfoodeng.2010.04.015

[55] Vitzrabin E, Edan Y. Adaptive thresholding with fusion using a RGBD sensor for red sweet-pepper detection. Biosystems Engineering. 2016;**146**:45-56. DOI: 10.1016/j.biosystemseng.2015.12.002

[56] Zhang H, Fritts JE, Goldman S a. Image segmentation evaluation: A survey of unsupervised methods. Computer Vision and Image Understanding. 2008;**110**(2):260-280. DOI: 10.1016/j.cviu.2007.08.003

[57] COMMISSION INTERNATIONALE DE L'ECLAIRAGE. (2004). CIE Technical Report Colorimetry 15:2004 3rd Edition.

[58] Zheng H, Lu H. A least-squares support vector machine (LS-SVM) based on fractal analysis and CIELab parameters for the detection of browning degree on mango (Mangifera Indica L.). Computers and Electronics in Agriculture. 2012;**83**:47-51. DOI: 10.1016/j.compag.2012.01.012

[59] García-Marino M, Escudero-Gilete ML, Heredia FJ, Escribano-Bailón MT, Rivas-Gonzalo JC. Color-copigmentation study by tristimulus colorimetry (CIELAB) in red wines obtained from Tempranillo and Graciano varieties. Food Research International. 2013;**51**(1):123-131. DOI: 10.1016/j.foodres.2012.11.035

[60] Liang Z, Sang M, Fan P, Wu B, Wang L, Yang S, Li S. CIELAB coordinates in response to berry skin anthocyanins and their composition in Vitis. Journal of Food Science. 2011;**76**(3):C490-C497. DOI: 10.1111/j.1750-3841.2011.02095.x

[61] Gallez L, Marconi A, Tourn E, González-miret M.L., Heredia F.J. Colour of Honeys from the SouthWestern Pampas Region: Relationship between the Pfund Color Scale and CIELAB Coordinates, In: Color in Food. Technological and Psychophysical Aspects. 2012, pp. 133-141, CRC Press Inc. ISBN: 978-1-4398-7693-0.

[62] Esparza I, Santamaría C, Calvo I, Fernández JM. Significance of CIELAB parameters in the routine analysis of red wines Relevancia de los parámetros CIELAB en el análisis de rutina de vinos tintos. CyTA - Journal of Food. 2009;**7**(3):189-199. DOI: 10.1080/19476330903068001

[63] van Roy J, Keresztes JC, Wouters N, De Ketelaere B, Saeys W. Measuring colour of vine tomatoes using hyperspectral imaging. Postharvest Biology and Technology. 2017;**129**: 79-89. DOI: 10.1016/j.postharvbio.2017.03.006

[64] Ibiza VP, Cañizares J, Nuez F. EcoTILLING in capsicum species: Searching for new virus resistances. BMC Genomics. 2010;**11**(1):631. DOI: 10.1186/1471-2164-11-631

[65] Sarpras M, Gaur R, Sharma V, Chhapekar SS, Das J, Kumar A, et al. Comparative analysis of fruit metabolites and pungency candidate genes expression between Bhut jolokia and other capsicum species. PLoS One. 2016;**11**(12):1-19. DOI: 10.1371/journal.pone.0167791

[66] Ryu WK, Kim HW, Kim GD, Rhee HI. Rapid determination of capsaicinoids by colorimetric method. Journal of Food and Drug Analysis. 2016:2-7. DOI: 10.1016/j.jfda.2016.11.007

[67] Stewart C, Mazourek M, Stellari GM, O'Connell M, Jahn M. Genetic control of pungency in C. Chinense via the Pun1 locus. Journal of Experimental Botany. 2007;**58**(5):979-991. DOI: 10.1093/jxb/erl243

[68] Das S, Teja KC, Duary B, Agrawal PK, Bhattacharya SS. Impact of nutrient management, soil type and location on the accumulation of capsaicin in Capsicum Chinense (Jacq.): One of the hottest chili in the world. Scientia Horticulturae. 2016;**213**:354-366. DOI: 10.1016/j.scienta.2016.10.041

[69] Duelund L, Mouritsen OG. Contents of capsaicinoids in chillies grown in Denmark. Food Chemistry. 2017;**221**:913-918. DOI: 10.1016/j.foodchem.2016.11.074

[70] Cisneros-Pineda O, Torres-Tapia LW, Gutiérrez-Pacheco LC, Contreras-Martín F, González-Estrada T, Peraza-Sánchez SR. Capsaicinoids quantification in chili peppers cultivated in the state of Yucatan, Mexico. Food Chemistry. 2007;**104**(4):1755-1760. DOI: 10.1016/j.foodchem.2006.10.076

[71] Powis TG, Murrieta EG, Lesure R, Bravo RL, Grivetti L, Kucera H, Gaikwad NW. Prehispanic use of chili peppers in Chiapas, Mexico. PLoS One. 2013;**8**(11). DOI: 10.1371/journal.pone.0079013

[72] Gutiérrez-Carbajal MG, Monforte-González M, de Miranda-Ham ML, Godoy-Hernández G, Vázquez-Flota F. Induction of capsaicinoid synthesis in Capsicum Chinense cell cultures by salicylic acid or methyl jasmonate. Biologia Plantarum. 2010;**54**(3):430-434. DOI: 10.1007/s10535-010-0078-z

[73] Giuffrida D, Zoccali M, Giofrè SV, Dugo P, Mondello L. Apocarotenoids determination in Capsicum Chinense Jacq. cv. Habanero, by supercritical fluid chromatography-triple-quadrupole/mass spectrometry. Food Chemistry. 2017;**231**:316-323. DOI: 10.1016/j.foodchem.2017.03.145

Chapter 5

A Proposal of Color Image Processing Applications for Education

Hiroshi Kamada, Tomohisa Ishikawa and
Keitaro Yoshikawa

Additional information is available at the end of the chapter

http://dx.doi.org/10.5772/intechopen.71001

Abstract

There are two main problems in the present style of education in which there are one teacher and many students in a class. The first problem is to improve communication between one teacher and many students in classes. The second problem is to realize personal education using IT systems. To solve the above two problems, we propose color image processing applications for education in this chapter. To improve communication between one teacher and many students, we realized automatic response analyzer that counts automatically students' answers using color cards raised by students. The system is an easy-to-use simple system, which only consists of a PC, web cameras, and color cards. The recognition rate was 98% in brighter class and 93% in darker class. To realize automatic personal education, we realized two interactive IT systems using color image recognition. First, we realized color learning system with color analysis functions. Using the system, students can learn color distribution in 3D color space for their selected images. Second, we realized visual programming system that judges automatically the correctness of the graphical image output by student program. If the student's output image is not correct, the system can point out the wrong part in the output image.

Keywords: color image processing, interactive communication, interactive learning, automatic response analyzer, color learning, programming learning

1. Introduction

Education is the fundamental and important activity to maintain the society. Ideal style of education is personal one teacher to one student education. However, main present style of education is one teacher to many students education in classes. Therefore, there are two problems in education.

First problem is to improve communication between one teacher and many students in classes. In the large classroom, one teacher needs to progress the lesson systematically for many students. In addition, it is necessary to grasp the degree of comprehension of the lesson for all students. For example, the most common way to grasp the status of many students is to ask questions or comments by raising hands. However, there are few students who raised hands and actively speak in the large classrooms. There is a need for a way to grasp the state of each student. To improve communication between one teacher and many students in classes, there are some conventional communication systems, which consist of a teacher's terminal and students' terminals. A teacher asks students using a teacher's terminal, and students answer to the teacher using students' terminals. However, conventional communication systems containing the student IT terminals disturb face-to-face communication, which is necessary for education in classes, and the systems are expensive in both initial cost and running cost.

Second problem is to realize personal education using IT systems. To solve the second problem, there are conventional e-learning systems. However, in conventional e-learning systems, automatic judging true or false questions are multiple-choice questions or perfect text-matching questions. Conventional system can treat only simple knowledge problems of which answer styles are multiple choice or perfect text matching, for example, not including free-style programming input, which is necessary for students to learn how to write their original programs using programming languages. Therefore, using conventional e-learning systems, students can cultivate basic abilities, but students cannot cultivate application abilities.

2. The research methods

The function to recognize each student status exists in personal one teacher to one student education, but the function doesn't in conventional systems. We decided to realize the function to recognize each student status in the IT system, because the students can learn as well as in personal one teacher to one student education using the IT system with the function in one teacher to many students education in classes or with no teacher. The representative function of recognizing each student status is the visual recognition function. Color images are important for communication and understanding in education area. To solve the above two conventional problems, we proposed the visual recognition function in IT system using color image recognition technologies. We proposed and realized the color image processing applications for education to solve conventional problems, and we verified the applications by experiments.

We describe the color image processing applications for education in Sections 3–5. With respect to the first problem to improve communication between one teacher and many students in classes, we describe the automatic response analyzer in classroom using image processing and cards in Section 3. With respect to second problem to realize personal education using IT systems, we focus on color learning area and graphics programming learning area, in which image processing is useful. We describe the color learning system with color analysis function in Section 4, and we describe the visual interactive programming learning system using image processing in Section 5. The system can judge students' programs output graphics answers right or wrong automatically.

3. The automatic response analyzer in classroom using image processing and cards

There are some conventional communication systems that consist of a teacher's terminal and students' terminals. As students' terminals, the dedicated miniature terminals [1–3], the dedicated digital pens [4], the student's mobile phones [5], and the student PC terminals [6, 7] are used. Although dedicated miniature terminals and dedicated digital pens have the advantage of not restricting the work space of students, management costs arise due to the risk of loss. For students' mobile phones, it is necessary for the students to pay communication packet cost. In a system using a PC terminal for students, it is necessary to make an investment to convert the whole classroom into IT classroom. Conventional communication systems containing the student IT terminals are expensive in both initial cost and running cost, and the systems disturb face-to-face communication, which is necessary for education in classes. Also, there was an attempt to grasp the status of many students by using qCards: a sheet of paper that contains a printed code, similar to a QR code, encoding their student IDs [8]. However, qCards are big. So it is inconvenient to carry, and we cannot do it easily.

3.1. The system architecture

We developed a simple system consisting of a PC and web cameras [9]. It is based on the method [10] of distributing multiple colored cards to students and letting the teacher to question the color card corresponding to the answer. The PC automatically compiles the card corresponding to the answer that the student cited. **Figure 1** shows the configuration of the system. We have mounted web cameras on the tripod and developed the connecting clasp. Since the two web cameras on the tripod are adjacent and high, there is no uncaptured area in front of area between two web cameras, and rear students' cards can be captured by the web

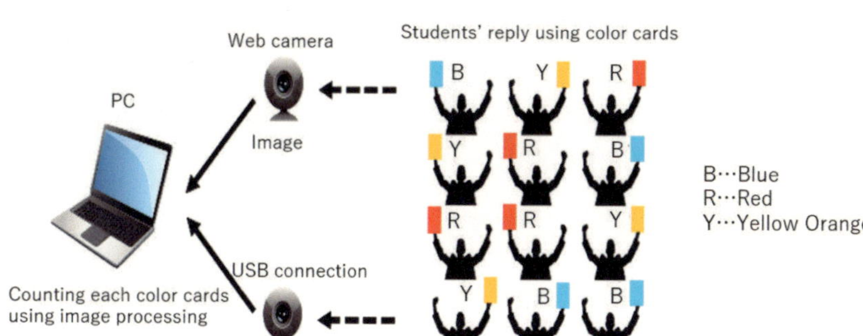

Figure 1. The configuration of the system.

cameras and students can be asked to raise color cards for questions from teachers. We show the color card's layout with A5 size in **Figure 2**. The color cards are made of two irreflexive papers that are fluorescent colored paper and black paper.

The process of the system consists of image acquisition and image processing. The image processing is divided into five stages. First stage is that we change the input image into the grayscale image. Second stage is that we change it into the binarized image. Third stage is that we extracted the area of the color portion from the binarized image. Forth stage is that we performed rectangle determination from the extracted area. Final stage is that we performed color determination after the rectangle determination. According to this system, compared to the conventional system, there is an advantage that the introduction cost is low, the maintenance is easy, and the usage method is simple.

3.2. Evaluation of the system

We experimented using eight candidate fluorescent colored papers shown in **Figure 3**. We show the specification of the classroom used for the experiment in **Table 1**. There are two lighting conditions. They are the bright lighting condition and the dark lighting condition. The bright lighting condition is that only the teaching side lights are turned off. The dark lighting condition is that the front side lights were also turned off.

At the first, we experimented using cards put on the desks and two web cameras. We show experimental results in **Table 2**.

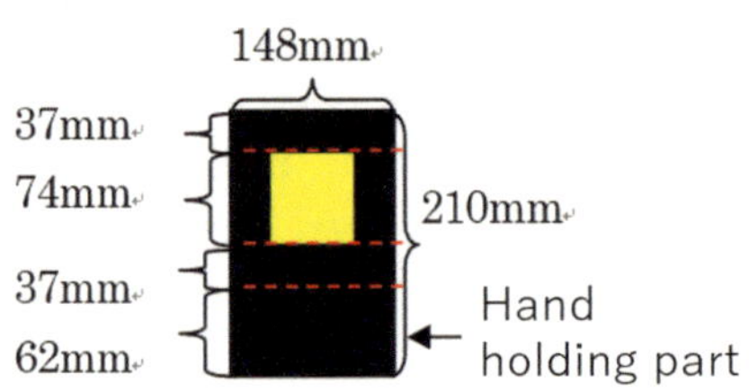

Figure 2. The color card's layout.

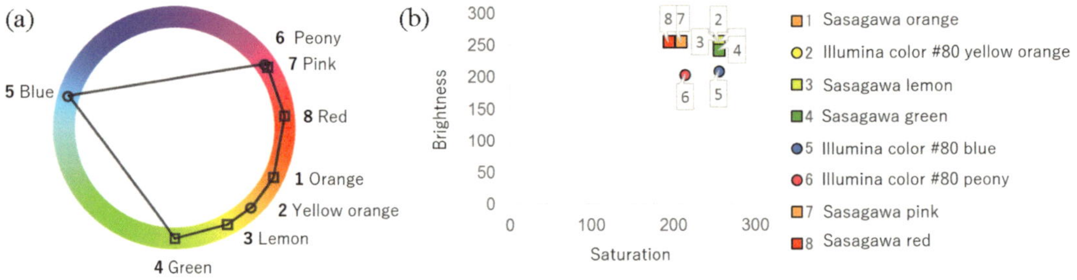

Figure 3. Color values of color cards. (a) Hue. (b) Saturation and brightness.

Width (m)	Depth (m)	Number of desks in width direction	Number of desks in depth direction	Number of seats in a desk
14.5	20.7	5	16	3

Table 1. The classroom specification.

Card name	Hue	Bright lightning	Dark lightning
Sasagawa orange	27	100%	100%
Illumina color #80 yellow orange	46	100%	100%
Sasagawa lemon	61	100%	100%
Sasagawa green	90	100%	100%
Illumina color #80 blue	197	100%	88%
Illumina color #80 peony	325	100%	89%
Sasagawa pink	327	100%	99%
Sasagawa red	354	100%	100%

Table 2. Experimental results.

We selected five color cards shown in **Figure 4** using **Table 2**. The layout of the classroom and classroom areas captured using three web cameras is shown in **Figure 5**. Image input device using three web cameras is shown in **Figure 6**. Card recognition rate depending number of web cameras is shown in **Table 3**. Questionnaire about usefulness of the system is shown in **Table 4**.

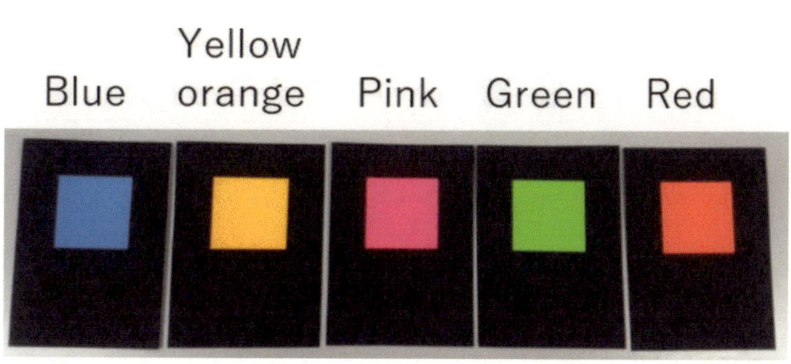

Figure 4. Color cards.

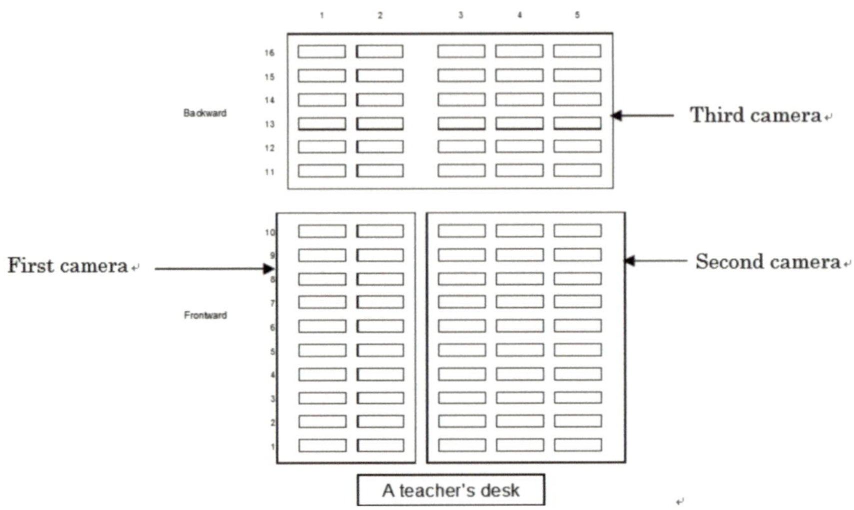

Figure 5. Classroom areas captured using three web cameras.

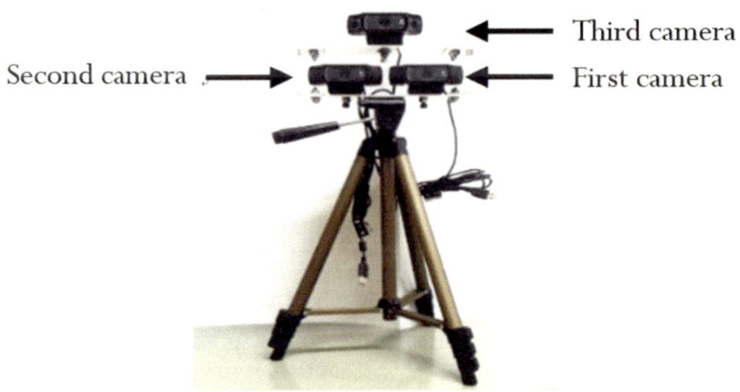

Figure 6. Image input device using three web cameras.

Web camera	Bright lightning		Dark lightning	
Two web cameras	97%	99%	81%	92%
		94%		66%
Three web cameras	98%	99%	93%	92%
		97%		95%

Table 3. Card recognition rate depending on the number of web cameras.

	Be useful	Neither	Not useful
A.D. 2015	44%	30%	26%
A.D. 2016	51%	31%	18%

Table 4. Questionnaire about usefulness of the system.

4. The color learning system with color analysis function

Since colors are used in various scenes of everyday life, effective learnings about colors are required. Conventional color learning uses analog methods such as books, color samples, and color charts, and efficient learning has not been done [11–13]. Therefore, it is necessary to develop a system that can efficiently learn colors on computers.

4.1. The system architecture

We developed a color learning system with color analysis function that enables efficient color learning digitally in previous research [14–20]. As shown in **Figure 7**, the system can be classified into three systems: a color learning system, an illusion simulation system, and a color analysis function.

The learning items of the color learning system are shown in **Table 5**. Among them, the most characteristic learning item is (9) color space (three dimensions). In the learning item (9), hue, saturation, and lightness can be three dimensionally simulated. The screen of the color space is shown in **Figure 8**. The students can move the color space three dimensionally with the mouse. By pressing the button on the left side of the screen, it is possible to switch display and nondisplay of hue, saturation, and brightness. Through this simulation, it is possible to visually understand the relationship between hue, saturation, and lightness.

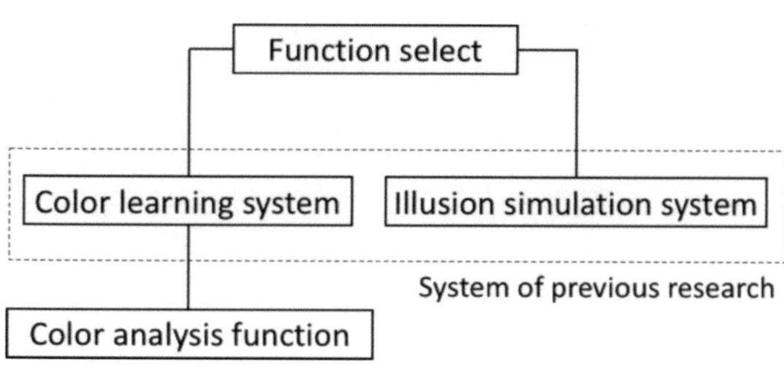

Figure 7. Configuration of the system.

	Learning items
(1)	Color information by color chart
(2)	Tone color
(3)	Tone name
(4)	PCCS tone diagram
(5)	Lightness, Saturation
(6)	PCCS hue circle
(7)	Hue
(8)	Color combination
(9)	Color space (three dimensions)

Table 5. Learning items of the color learning system.

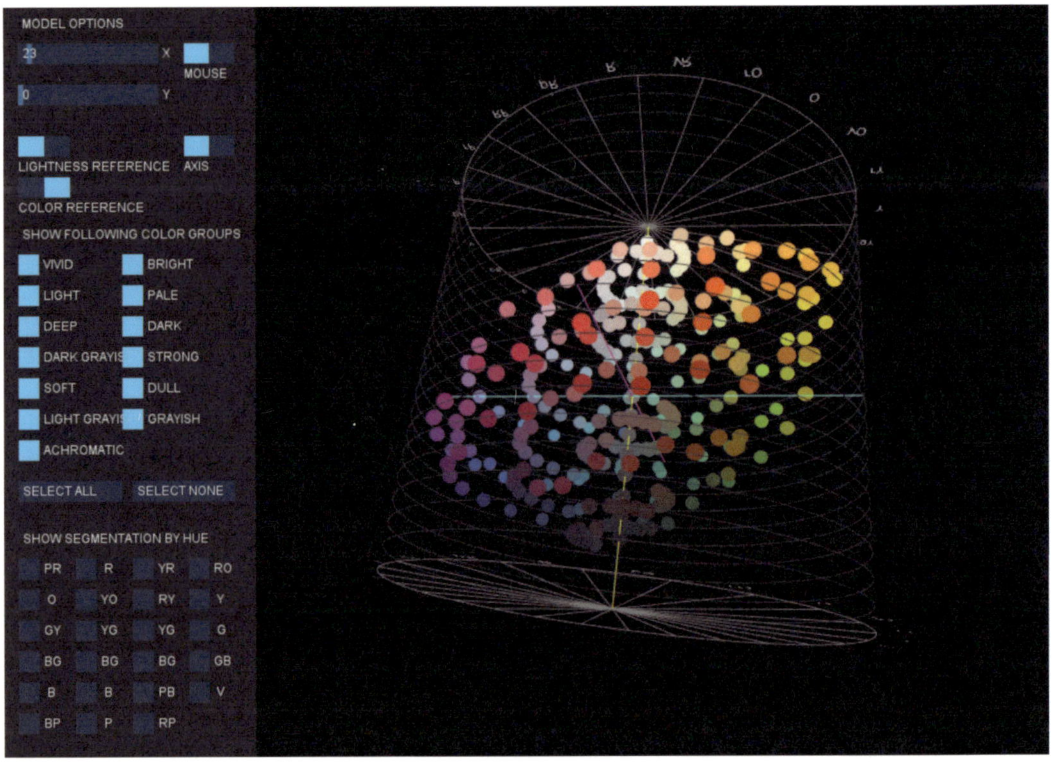

Figure 8. An example of color space.

The illusion simulation system is a system that simulates an illusion phenomenon by freely changing parameters such as color and motion [17]. The system can simulate 15 kinds of illusion phenomena from five categories of geometrical illusion, light illusion, illusion of color, illusion of shape, and motion illusion.

http://dx.doi.org/10.5772/intechopen.71001

An example of a simulation screen of Ebbinghaus illusion is shown in **Figure 9**. The size, the number, and the position of the white ball around the yellow size ball having a constant size are displayed in two ways with different left and right. Ebbinghaus illusion is the illusion that a ball surrounded by a large sphere is small, and a ball surrounded by a small sphere looks big. In **Figure 9**, the ball on the left is larger than the ball on the right.

The color learning system including illusion simulation is capable of interactive learning, but it is passive learning in terms of answering the presented problem. In addition, there is no function that can easily check and reproduce any color you enter. Therefore, a learning system that makes active learning by visually converting input arbitrary color information, confirms and reproduces color information by learners, and leads to deep understanding of colors is required.

We developed a color analysis function as a system that visualizes color information and enables active learning. The color analysis function is a function that can visually understand the color information of the color by visualizing the hue, saturation, and brightness of the pixel three dimensionally. When you click the mouse on the displayed image, you get the hue, saturation, and brightness of the pixel at the clicked position. Then, a sphere is displayed in the coordinates corresponding to the hue, saturation, and brightness acquired in the 3D model of **Figure 10**, where the circumference is hue, the x axis and y axis are saturation, and the z axis is lightness. Every time the mouse is clicked, a sphere is added, and multiple spheres are displayed. Also, at the bottom of the displayed image, the hue, saturation, and brightness of the pixel selected with the mouse are displayed.

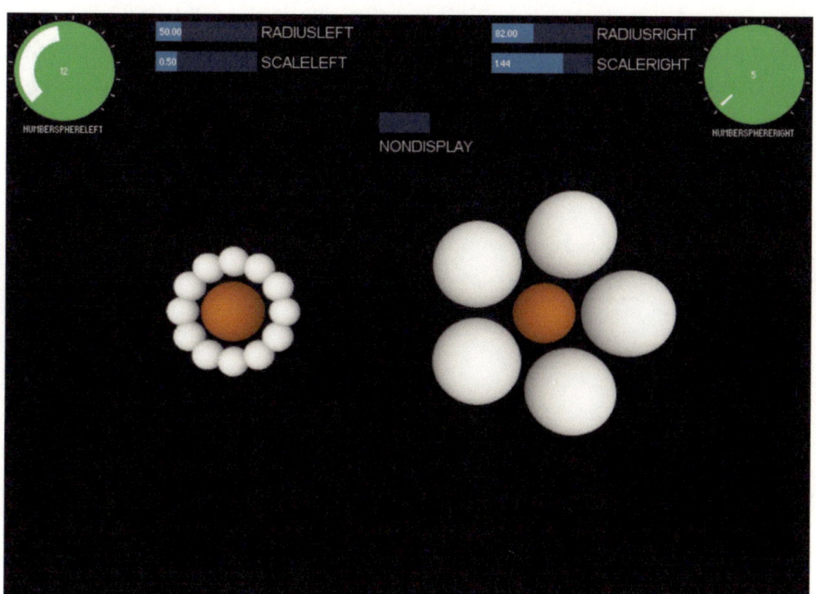

Figure 9. Ebbinghaus illusion simulation example.

Figure 10. 3D model for color analysis.

An application example of the color analysis function is shown in **Figure 11**. **Figure 11(a)** show a test image. The middle oblong rectangle is the same gray from the left end to the right end. On the other hand, the background continuously changes from black at the left end to white at the right end. The mouse pointer is placed almost at the left end of the middle oblong rectangle. The HSV value of the pixel at this position is displayed at the bottom. In **Figure 11(b)**, the HSV value of the pixel is shown on the color analysis function 3D model. Even if you move the mouse pointer rightward in a horizontally long rectangle, the points on the 3D model do not move. This confirms that the middle oblong rectangle is a single color. For human eyes, the horizontally elongated rectangle in the test image appears bright at the left end and dark at the right end, but it can be realized that this is an effect of light contrast against the background.

Figure 11. 3D model for color analysis function. (a) Test image. (b) 3D display of image pixel.

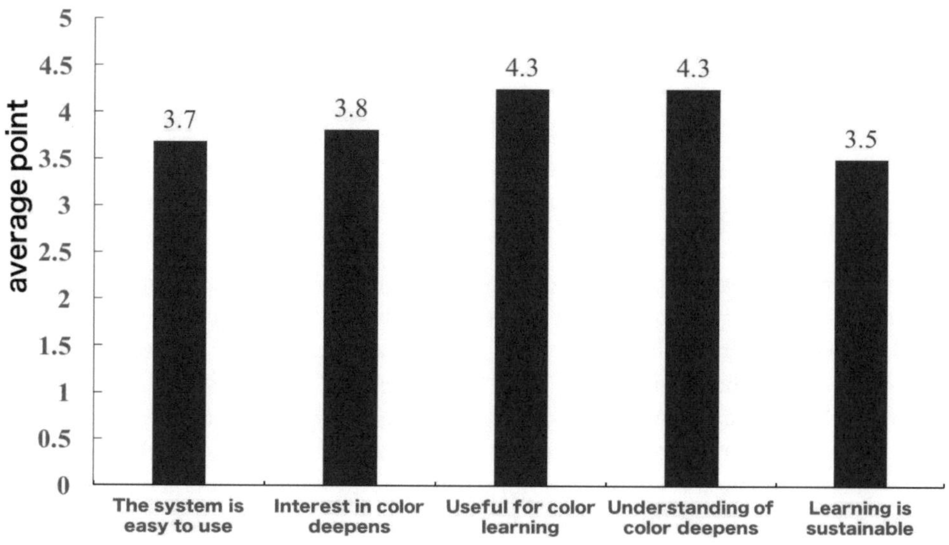

Figure 12. Questionnaire.

4.2. Evaluation of the system

The subjects were given a color learning system and an illusion simulation system, the purpose and usage of the color analysis function and a questionnaire were carried out. In the questionnaire, the comprehensive evaluation of each system was made to answer by a five grade evaluation for each question item. The questionnaire result is shown in **Figure 12**.

5. The visual interactive programming learning system using image processing

Nowadays, since many software must be made constantly to maintain IT society, we must bring up many programmers constantly. Therefore, there are many persons learning programming. However, there are fewer persons to master programming skills, because it is difficult for persons to keep high interest and high motivation to write complicate and long programs by themselves. Most of conventional learning programming processes are the processes of learning program language grammar, which are necessary but are not efficient for students to acquire the skill to write complicate and long programs. There are two different types of conventional programming educations reports [21, 22]. In the first report [21], "a self-study supporting system which is an automatic fill-in-the-blank problem generator" is proposed. However, the system is not suitable for educations to write complicate and long programs. In the second report [22], "programming experiments" is proposed. In the education, game or other interesting program functions are implemented by students. However, the students are educated only in the classes at school. There are no conventional programming educations or learning systems for students to keep high interest and motivation to write complicate and long programs by themselves.

5.1. The system architecture

To solve the program, we propose the visual interactive programming learning system using image processing. The system requires the students to write the programs that output the indicated figures, and the system compares the correct figures with the student's program output figures using image processing, and the system judges that students' programs are correct or not. The system concept is shown in **Figure 13**. First, question is displayed to the student to complete the program that outputs the indicated figure. Second, the student inputs the program to complete the whole program. Third, the system outputs the image file of the figure that the student program displays, and the system outputs the image file of the figure that the correct program displays. Fourth, the system compares the two images using image processing, and the system judges that students' programs are correct or not. Furthermore, the system may point out the different parts of two images. The system may be expected to be valid for students to keep high interest and motivation to write complicate and long programs by themselves.

The programs may display the figures by multiple steps. The programs may display moving images, too. Correspondingly, the system compares the multiple pair of images or a pair of

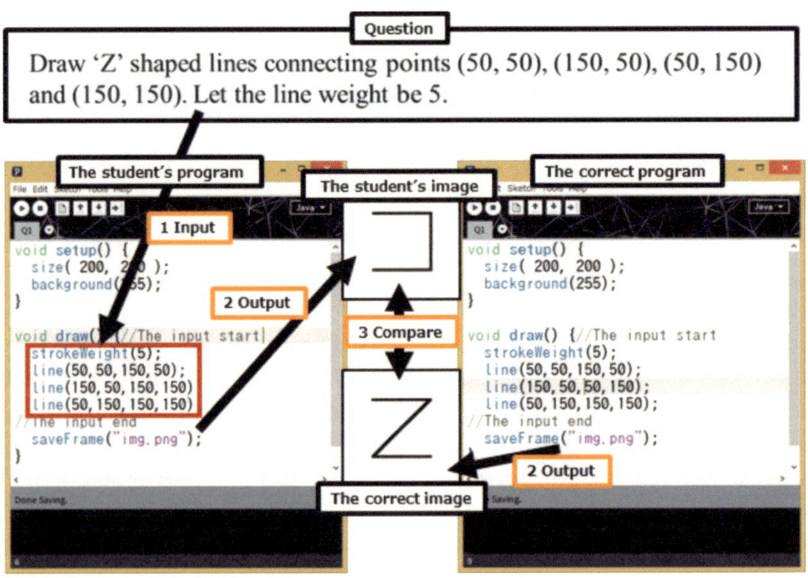

Figure 13. The system concept.

moving images using image processing, and the system judges that students' programs are correct or not. We will describe the three system functions.

Most simply, the system compares a pair of still images that student's program and the correct program display, and the system judges that students' programs are correct or not. The judgment using a pair of still images is shown in **Figure 14**. To answer the question, the student inputs the student's program into the input field at the input form. In **Figure 14**, there is the mistake about a circle position. The different parts between the two images that student's program and the correct program output are displayed in red. Example of a judgment using a pair of still images is shown in **Figure 15**. In this case, the correct image is "×", but the student's image is "+". In each images, the different part is displayed in red.

The system may find the wrong parts of student's programs using plural pairs of output images. The judgment process using plural pair of images is shown in **Figure 16**. There are plural input fields in the input form. The students input programs in the plural input fields. The completed program outputs the displayed images at the end of each input field. The system compares the student's program output image and the correct program output image at the end of each input field. Example of a judgment using plural pair of images is shown in **Figure 17**. In these examples, two images output at the end of the first input field are same, but two images output at the end of the second input field are different. Therefore, the student's program in second input field is wrong. The different parts in the two images are displayed in red.

Furthermore, the system can judge moving output images. The judgment using a pair of moving images is shown in **Figure 18**. The input field in the input form is in the program loop,

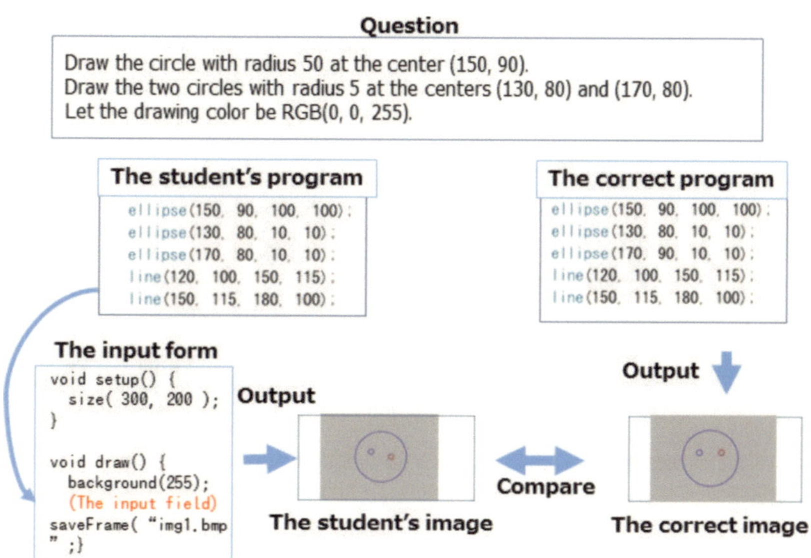

Figure 14. Judgment using a pair of still images.

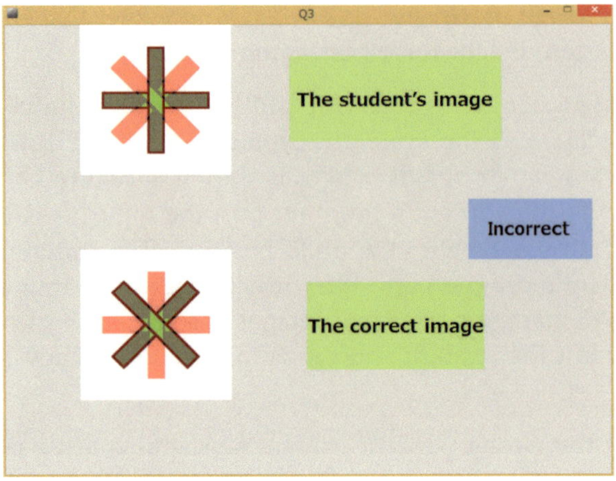

Figure 15. Example of a judgment using a pair of still images.

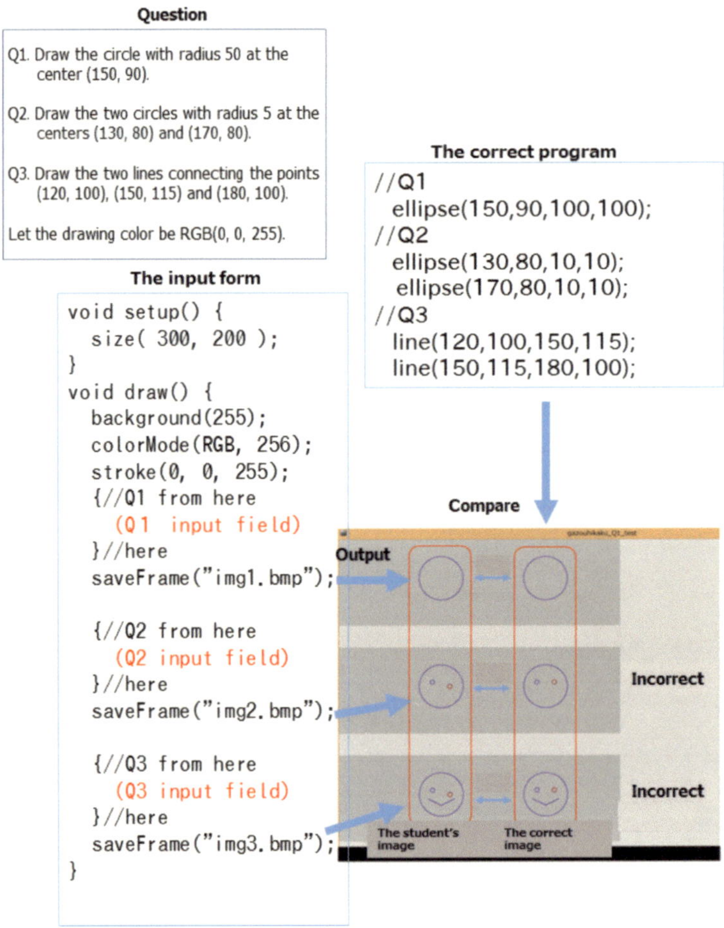

Figure 16. Judgment using plural pair of images.

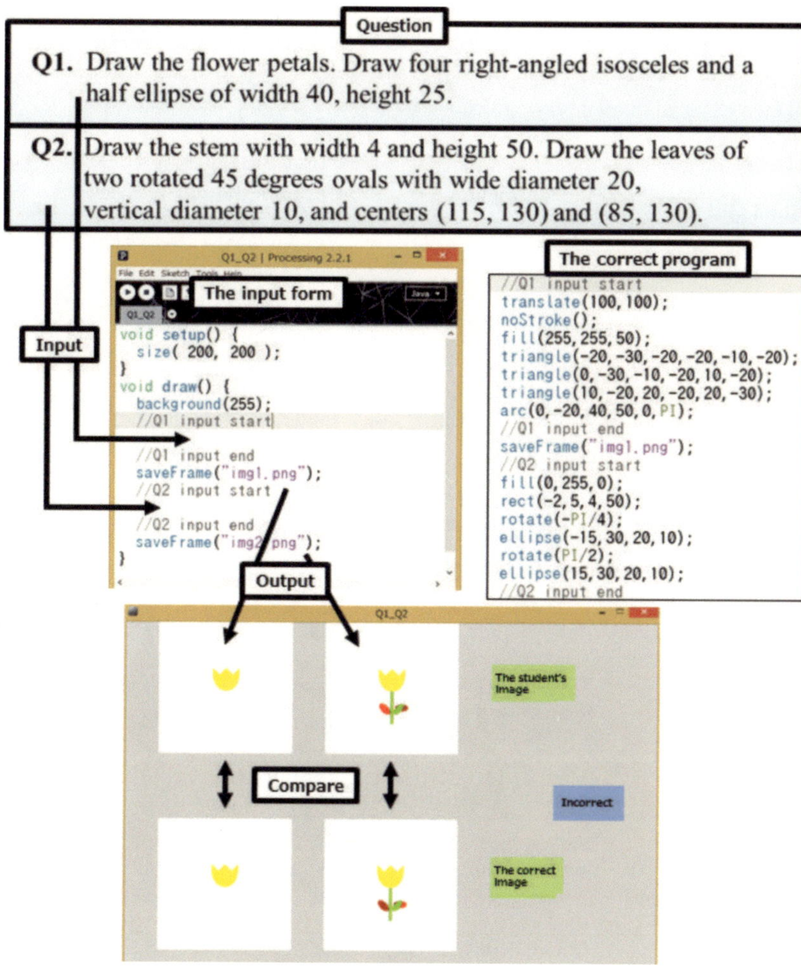

Figure 17. Example of a judgment using plural pair of images.

and at the end of the each program loop, the displayed image is output. The system compares the student's program output image and the correct program output image at the end of each loop. Example of a judgment using a pair of moving images is shown in **Figure 19**. Different parts of two images are displayed in red.

We implemented the function describing in this Section 5.1, and we developed the programming problems of which are about drawing point, line, color, figure, and animation, etc.

5.2. Evaluation of the system

We developed the system, and we evaluated the system using questionnaires. We requested the examinees to answer questionnaires that contained the following three questions.

Q1. Is it easy to use the system?

Q2. Do you feel that you can learn programming using the system?

Q3. Can you keep motivation for learning programming by the system?

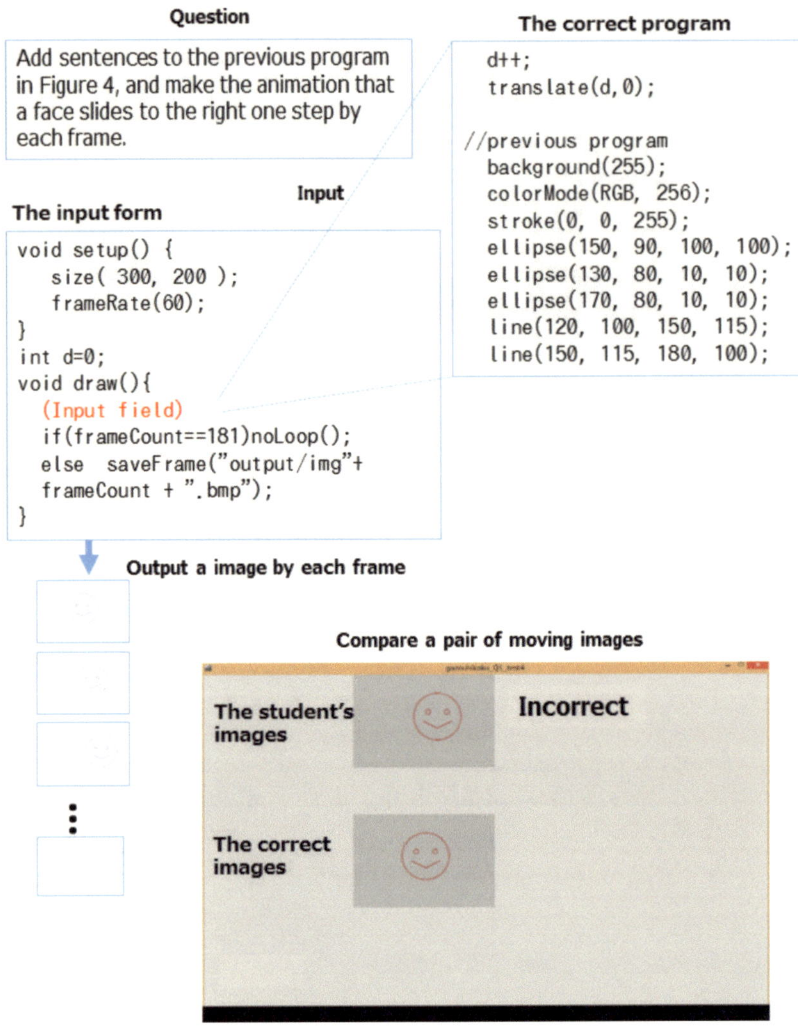

Figure 18. Judgment using a pair of moving images.

There are five kinds of answers, "strongly agree" with five points, "agree a little" with four points, "neither agree nor disagree" with three points, "disagree a little" with two points, "strongly disagree" with one point. The result of the first experiment is shown in **Table 6**. The examinees are 10 students of Kanazawa Institute of Technology. Average points of the questionnaires are 3.7, 4.1, and 3.8. Based on the questionnaires result, we think that the system concept is useful, but the improvement of the system is necessary for practical uses.

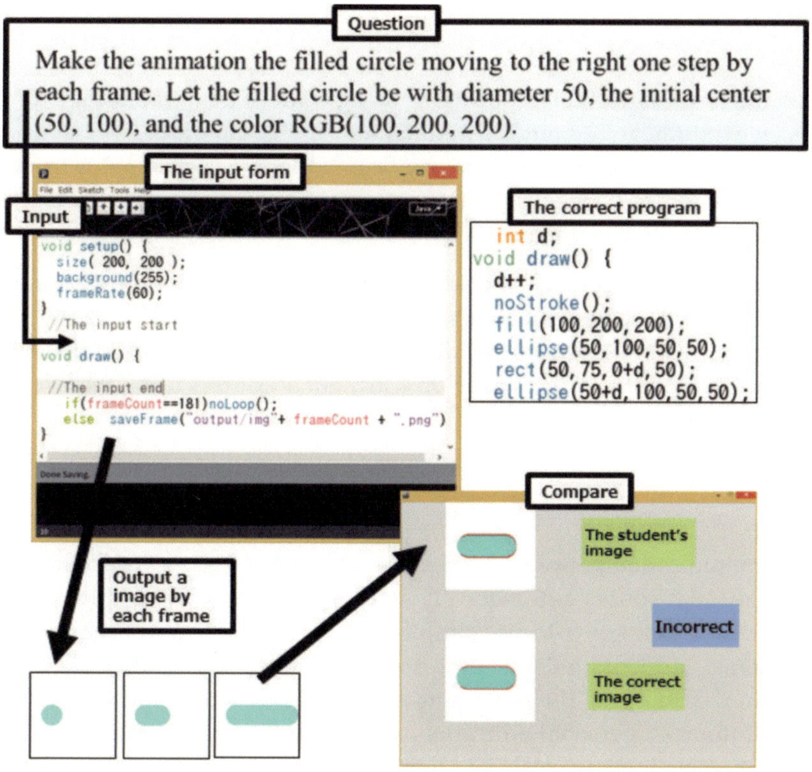

Figure 19. Example of a judgment using a pair of moving images.

Evaluation level (points)	Q1	Q2	Q3
Strongly agree (5)	2	2	1
Agree a little (4)	5	7	6
Neither agree nor disagree (3)	1	1	3
Disagree a little (2)	2	0	0
Strongly disagree (1)	0	0	0
Average points	**3.7**	**4.1**	**3.8**

Table 6. Questionnaire results (number of respondents).

6. Consideration

We proposed the three education applications using color image processing. We verified the applications by experiments. About the automatic response analyzer, we have already used the application in our classes, and we have confirmed that the system helps teachers to manage classes more smoothly. Next study is to develop various functions that help class management to achieve more educational effects.

We have developed prototypes of the color learning system and the programming learning system, and we verified the new functions. However, we have not used the two applications in actual educational classes, because we do not have developed educational courses using the two applications. Next study is to develop educational courses using the two applications.

7. Conclusion

To improve communication between one teacher and many students, we realized automatic response analyzer that counts automatically students' answers using color cards raised by students. The system is an easy-to-use simple system, which only consists of a PC, two or three web cameras, and five kinds of color cards. The recognition rate was 98% in brighter class and 93% in darker class. The system made the teacher's educational management easier. To realize automatic personal education, we realized two interactive IT systems using color image processing. First, we realized color learning system with color analysis functions. Using the system, students can learn color distribution in 3D color space for his or her selected images. Second, we realized visual programming system that judges automatically the correctness of the graphical image output by student program. If the student's output image is not correct, the system can point out wrong part in the output image. We proposed above the three systems, and we verified the systems by experiments.

Acknowledgements

This work was supported by JSPS KAKENHI grant number 15 K01041. We thank students and former students who have cooperated with researches in this paper.

Author details

Hiroshi Kamada*, Tomohisa Ishikawa and Keitaro Yoshikawa

*Address all correspondence to: kamada@neptune.kanazawa-it.ac.jp

Kanazawa Institute of Technology, Hakusan, Japan

References

[1] TERADA.LENON Co. Ltd. LENON System 2016 [Internet]. Available from: http://www.t-lenon.com [Accessed: Sep 5, 2017]

[2] IC Brains Co. Ltd. Socrates System 2015 [Internet]. Available from: http://www.icbrains.com/soctop.html [Accessed: Sep 5, 2017]

[3] Sugihara T, Miura M, Sakamoto Y, Endo S. Stage in the classroom: Proposal of interactive lesson using digital pen. Information Processing Society Research Report. 2009; **2009-HCI-133**(3):1-8

[4] Miura M, Kunifuji S, Shizuki B, Tanaka J. Real-world oriented interactive learning system based on digital pen devices and PDAs. IPSJ Journal. 2005;**46**(9):2300-2310

[5] Kunori N. Experiments on university multiplayer lecture using e-learning by mobile phone. Media Education Research. 2005;**1**(2):145-153

[6] Wingnet System. Computer Wing Co. Ltd 2012 [Internet]. Available from: http://www.cwg.co.jp/ [Accessed: Sep 5, 2017]

[7] Matsuuchi N, Shiba N, Yamaguchi T, Huziwara K. Practice and evaluation of interactive support system providing spontaneous active learning environment. Information Processing Society. 2008;**49**(10):3439-3449

[8] Cross A, Cutrell E, and Thies W. Low-cost audience polling using computer vision. In: Proceedings of ACM Symposium on User Interface Software and Technology (UIST); 2012

[9] Kamada H, Yamada K, Yoshikawa K. Automatic response analyzer in classroom using image processing and cards. Automatic response analyzer in classroom using image processing and cards. ICIC Express Letters Part B: Applications. 2016;**7**(8):1719-1725

[10] Suetake K. An improvement of education in university using education engineering. Kanagawa University Engineering Laboratory Report. 1989;**12**:23-38

[11] National Federation of Fashion Educators. Color Assessment Official Text Third Level. AFT Planning. 2008

[12] National Federation of Fashion Educators. Color Assessment Official Text Second Level. AFT Planning. 2009

[13] National Federation of Fashion Educators. Color Assessment Official Text First Level. AFT Planning. 2009

[14] Hiroshi Kamada. Color design learning system using computer image processing. Papers from the CIEC Academic Meeting. 2013;**4**:60-64

[15] Zambrano MM, Kamada H. Interactive education system using image processing and computer graphics. In: Reports of Mexico training program for the Strategic Global Partnership. Kanazawa Institute of Technology. 2013

[16] Kamada H. A color design learning system using computer image processing. ICIC Express Letters, Part B: Applications. 2014;**5**(1):88-94

[17] Watanabe Y, Miyazaki S. Color experience experience system through illusion simulation. [Kanazawa Institute of Technology graduation thesis]. 2014. p. 1-165

[18] Tomohisa Ishikawa, Hiroshi Kamada. Color learning system including illusion simulation. In: Proceedings of 2016 PC Conference. 2016. pp. 163-166

[19] Tomohisa Ishikawa, Hiroshi Kamada. Color learning system with color analysis function. Selected papers from the 2017 CIEC Academic Meeting. 2017;**8**:23-27

[20] Tomohisa Ishikawa, Hiroshi Kamada. Color learning system with color analysis function. In: Proceedings of 12th International Conference on Innovative Computing, Information and Control. 2017. pp. 176

[21] Hara K, Yan Y, Nakano H, Chino H, Kazuma T, He A. A fill-in-the-blank problem generator for C program beginner. IEICE Technical Report. 2015;**115**(127):37-42 /ET2015-29

[22] Imani J, Osana Y, Kikuchi M, Ito M, Ishihata H. A practical report on the programming in first year introductory education. IEICE Technical Report. 2015;**115**(285):43-48 / ET2015-51

Section 2

Image Processing

Chapter 6

Real-Time Video Analysis in Agriculture by Using LabVIEW Software

Abdullah Beyaz

Additional information is available at the end of the chapter

http://dx.doi.org/10.5772/intechopen.71236

Abstract

Technological developments help us to make our lives easier, such as video analysis systems. Because of that reason, these systems are commonly used in different fields. So, this kind of applications is rising these days. There are two important factors for widespread of this technology. The first factor is an increase of video capture performances at video processing units, and the second factor is decrease in prices of the video processing units. In this chapter, some general information is given about video processing technique and the information held by a LabVIEW application. As LabVIEW application, finger anthropometric measurements have been performed by utilizing the video analysis technique for agricultural machinery and instrument evolution. At the application, real-time anthropometric measurements have been done from instantly obtained video images. Real-time anthropometric measurements of video images were obtained from Ankara University Faculty of Agriculture students by using LabVIEW software. To check result accuracy, the obtained value was compared with the caliper measurements. After that differences between the results were examined statistically. In the performed study, all real-time video image values show that this system can be used effectively for real-time anthropometric measurements.

Keywords: LabVIEW, video analysis, agriculture, anthropometric measurements, ergonomics

1. Introduction

The human visual system is limited to a very narrow portion of the spectrum, so today image processing is used in wide variety of applications. By this way, it is possible to improve the visual appearance of images to a human. A vision system contains an electronic unit for image acquisition from devices like human vision system. There are a lot of computer vision systems in many areas to use as inspection systems to check the size of objects. Likewise, we can utilize them for biometric measurements. This chapter contains the comparison of two measurement methods that are commonly used for analysis processes. One of them is caliper measurement as experimental measurement, and the other one is video measurement system as an image processing application.

Nowadays, video analysis systems and applications are increasing with technological developments. These systems and applications are helping us to make our lives easier. There are two important factors effects widespread of this technology: the first factor is the superior performance of video capture devices, and the second factor is decreasing at prices of video processing units [1].

Several studies have been done in this field with different objects for developing more efficient systems. Nowadays, it is also developing efficiently. A lot of programming languages (C, C++, C#, Visual Basic, Delphi, etc.) and also software development platforms (MATLAB, LabVIEW, etc.) can be used for this aim. The image measurements are often used as a method for acquiring scientific data. This method generally requires well-defined features or structure, also edges or unique color, texture, or some combination of these factors. These types of measurements can be performed on all of the visual scenes or on individual features that are important for determining on the visual scenes [2]. The inspection steps for building a vision application can be seen in **Figure 1**.

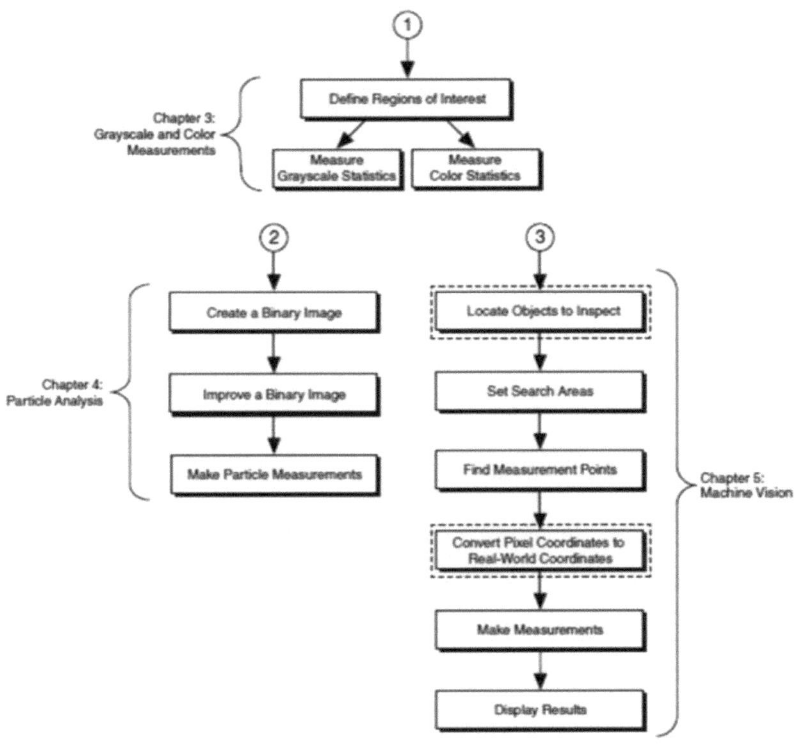

Figure 1. Inspection steps for building a vision application [3].

In this chapter, some general information is given about video processing technique. For this aim, a LabVIEW application is presented for explaining the technique. There are a lot of applications at literature in this field. For example, Szabo et al. [4], in their study, defined the color of objects by using LabVIEW 8.6, NI Vision Development Module, NI IMAQ for USB software. Then they have compared the results with the software developed by them, which is written in LabWindows CVI/ANSI C programming language. They also reported that developed system could be successfully used in robotics fields and object classification.

Pal Singh and Julka [5] in their research classified different sizes of screws from getting images by using LabVIEW Vision Assistant software. They have indicated that the software successfully classified screws.

Ravi Kumar et al. [6] in their research stated that determination of edges could be done successfully from real-time video images by using the software which is developed by them at LabVIEW platform.

Dilip and Bhagirathi [7] in their research claimed that defined coins could be determined by using a smart camera (NI-1742) and the software which is developed on LabVIEW 2010 Vision Assistant platform. They emphasize that the software identifies the coins successfully.

These researchers show that video processing at LabVIEW platform is reliable, fast, comfortable, and inexpensive. Because of these reasons, video processing techniques on the LabVIEW platform are explained in this chapter with an application from the ergonomic field for the aim of agricultural machinery and tool development. First of all, we have to know why we use ergonomics in agricultural machinery and tool development. The answer of this question is "Ergonomics interested in design and design efficiency for a human. Also, it means wherever design and design efficiency interact with human-environment-machine systems." So, ergonomic principles are essential for human-based systems. Knowing the anthropometric dimensions of the human physical structure is a prerequisite for the design of human-machine

Figure 2. Hand anthropometry application samples in agricultural tools and machines.

interaction systems. Working with the help of muscle force, machine, and tool compatibility is important regarding work efficiency [8].

Different forms of control elements can be found on agricultural machines such as a handle, lever, and buttons. So, dimensions of agricultural tools are essential for human anthropometry. These control element dimensions should be set in each way according to the anthropometric measurements and applied science data (**Figure 2**). When we look at this perspective, the issue is important for the development of agricultural machinery design and manufacturing [8].

In this application, video images were used for anthropometric measurements which were obtained from real-time video. For this intention, hand images are obtained from Ankara University, Faculty of Agriculture students.

2. Material and method

2.1. Material

2.1.1. Image processing languages

The appropriate programming language and programming platform are significant for the image analysis applications. Also in real-time imaging, we have to choose logical and accessible interfaces for the best applications that we need. Additionally, it means right hardware and software combinations. There are a lot of programming languages such as C, C++, C#, Java, Visual Basic, FORTRAN, BASIC, etc., to implement real-time color imaging systems. In many instances, they have adverse effects like you sustain to know deeply programming information for many image analysis applications. But the LabVIEW platform gives an opportunity for rapidly image processing applications than C and other programming languages for real-time color imaging systems. It is probably a good thing because of the secure programming benefits for agricultural engineers. Understanding the performance impact of various language features is difficult. There is no clear answer, for experimentation with the language compiler and performance of measurements. But LabVIEW tools can be helpful for obtaining the most useful measurement results.

2.1.2. LabVIEW

LabVIEW is a software development platform with a graphical interface that millions of technologists and scientists use. Rather than other software development platforms, LabVIEW offers comfortable and more visual platform for developing software. Ever since 1986, LabVIEW platform improved great graphical programming library, and now it offers great convenience in all areas where we use engineering data [9].

2.1.2.1. LabVIEW graphical interface and sample application

2.1.2.1.1. Front panel

LabVIEW software has two components for graphical programming as front panel and block diagram. The front panel is an interactive software part where the user can enter data in

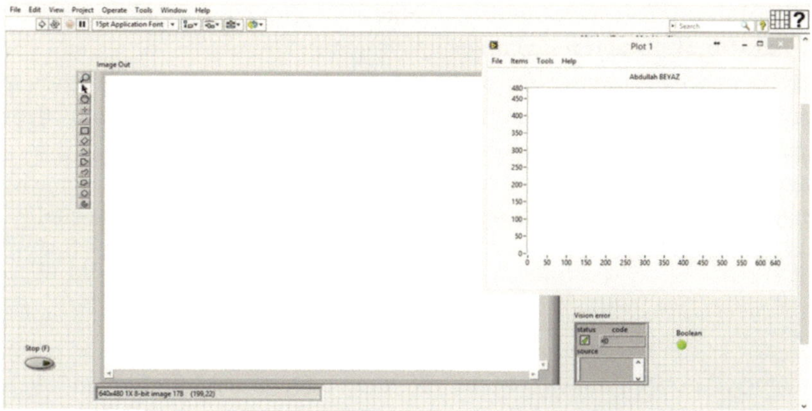

Figure 3. Front panel of sample application.

software and which has an interface area with different visual elements like indicators and graphics; it is also based on human-machine interaction. It means you can watch the screen output. By the visual elements placed in this field, the interface of needed software can be created in a short time. **Figure 3** shows the front panel of a sample application which has been developed by using LabVIEW.

2.1.2.1.2. Block diagram

The block diagram is the place where the actual programming operation was done. The source codes are run in this place which is created by using the links between virtual objects. Post-processing outputs can be seen on the front panel after software input and output processing. All virtual elements which have input and output ports can be seen as an icon. The block diagram is created by using this icon combination for software development. **Figure 4** shows the block diagram which was developed by using LabVIEW platform for the sample application.

2.1.3. Hardware and display issue

Understanding the hardware support for image accusation is fundamental of the analysis of real-time performance of the color image measurement systems. Some computers contain special

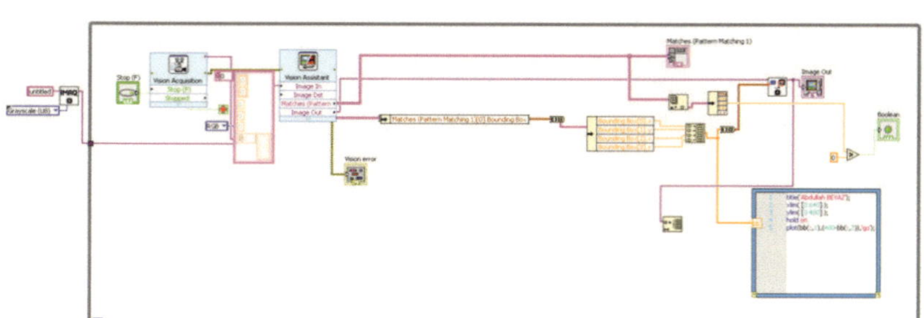

Figure 4. Sample application block diagram.

hardware for real-time image measurement applications with high-performance processors with the optimum structural support for complex imaging measurements. The technological developments give us a chance for buying low-cost pixel processors for real-time imaging applications. But commonly personal computers are used for color image measurement systems. There are many structural issues relating to real-time performance like memory capacity, memory access times, storage device speed, display hardware performance, and storage capacity. Real-time design of imaging measurement systems contains the necessary decisions like quality or working speed versus resolution [10].

2.1.3.1. Cameras

Recent technological developments reduced the complexity of the imaging devices, and also the cost of these color devices decreased. So, the popularity of these devices is still increasing for the consumers. Single-sensor digital cameras are used today in research activities in the area of color image acquisition, processing, and storage. Single-sensor camera image processing methods are getting more significant with the developments of digital camera-based applications.

LabVIEW supports lots of cameras which have a different communication protocol. Some of them are wireless, and some of them have wire connections. These cameras can be classified as bus type, scan type, and color type. Bus-type cameras can be classified as GigE, Usb3, Camera Link, IEEE 1394, IEEE 1394b, IP camera, and Parallel Digital. Scan-type cameras can be classified as area scan and line scan cameras. Also, color-type cameras can be classified as color and monochrome cameras. All supported cameras can be found easily on National Instruments camera network web page (**Figure 5**).

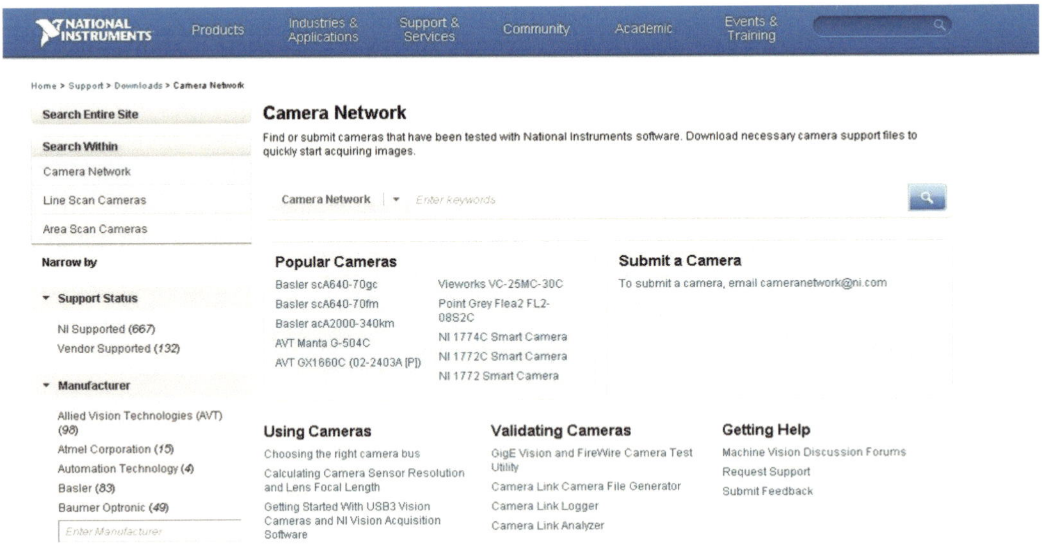

Figure 5. National instruments camera network webpage.

http://dx.doi.org/10.5772/intechopen.71236

2.1.3.2. DAQ devices

In this application, we did not use any DAQ devices for measuring or sending analog and digital signals, but some applications require DAQ devices for controlling or measuring processes [10].

DAQ devices have an interface between a PC and the real world for outside signals. The first function of DAQ devices is taking analog and digital signals for processing, and the second function is sending analog and digital outputs for controlling a physical unite. There are many ways for measuring different types of signals that we required. A sensor or measurement device sends analog and digital signals of a physical phenomenon for measurements like electrical voltage. You can also send analog and digital output signals for controlling a physical unit. For this reason, it is important to understand analog and digital signal types for controlling their actions. Based on your analog and digital signals of your application, you can decide which DAQ device you need. There are a lot of functions of DAQ devices, for example, you can measure analog-digital input and output. Also, you can generate analog and digital input and outputs [10].

There are different types of DAQ devices for one or multiple functions which support your application. You can find DAQ devices, according to input or output channels of your application, but you have to decide the price of the needed DAQ device. This decision can be done from the web page of supported NI products (**Figure 6**).

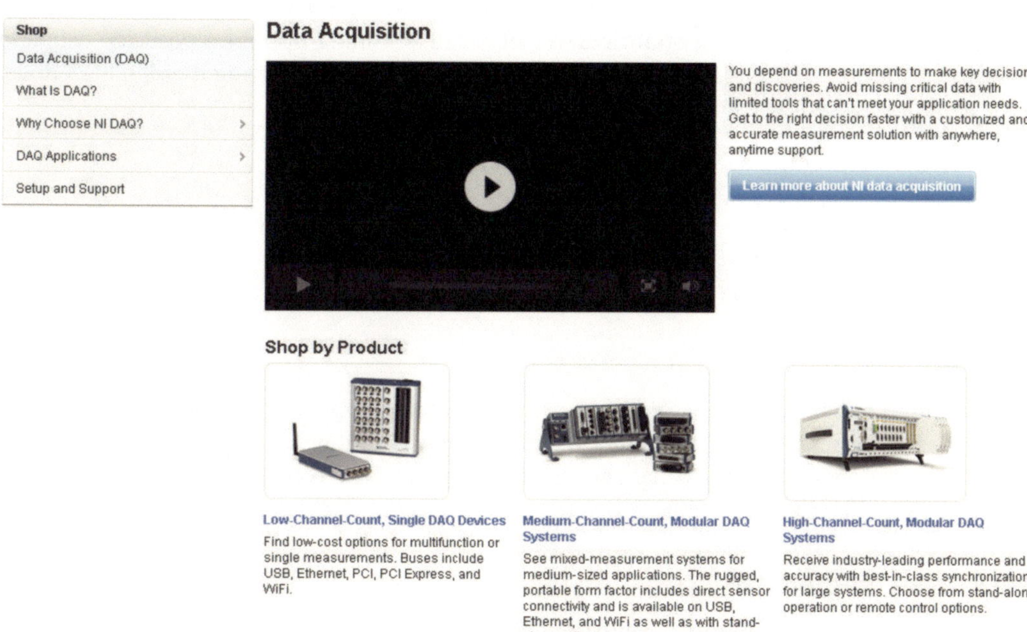

Figure 6. National instruments DAQ devices webpage.

2.2. Method

2.2.1. *Measurement types*

2.2.1.1. *Surface area measurement*

Surface areas describe the boundaries between regions. A surface area analysis of the different components of objects gives valuable data with the prediction relations. Thus, a scientist uses different methods for assessing areas of targets that they concern, and one of them is an image processing and analysis technique [10].

2.2.1.2. *Length measurement*

Length measurement is usually applied to objects that have one long dimension in comparison to others. For small objects like buttons or control arm's length as machine parts, measurement can be performed by applying image analysis and processing technique. Photographic enlargement of the dimensions helps us to assess the length of agricultural products easily [10].

2.2.1.3. *Determining number*

In the field of machine vision, blob detection refers to visual modules that are targeted at detecting points or regions in the image that are either brighter or more colored than the smothering. Also, determination of the numbers of objects is important for agricultural applications. We can define the specific gravity of seeds, thousand grain weight, a number of droplets in spraying, etc., for learning biological material properties. These attributes also can be applied for agricultural machine and tool development. For instance, the numbers of the droplets on a leaf in pesticide application are an important parameter for determining optimum pesticide usage in agricultural products, because pesticides affect environment and employing them in high doses cause air, land, and water contamination [10].

2.2.1.4. *Color measurements*

RGB and CIELab are widely used as standard spaces for comparing colors. A set of primary colors, such as the RGB primaries, defines a color triangle; only colors within this triangle can be reproduced by mixing the primary colors [10].

2.2.1.5. *Determining location*

There are several definitions of location and pixels in an ordinary image. For instance, the x and y coordinates of the midpoint of a feature can be determined simply by the minimum and maximum limits of the pixels. Normally, starting from the top left corner of the array, the pixel addresses them. The global coordinate system gives us info about real pixel dimension values [10].

2.2.1.6. *Neighbor relationships*

Local coordinates and individual features of the pixels are important for some applications, and neighbor pairs are the easiest way to identify differences between the products. The histogram of the distribution of these features gives the answers [10].

2.2.1.7. *Perimeter measurements*

The perimeter of a feature is a well-defined and familiar geometrical parameter. Measuring a numerical value that identifies the object properties can be easily used to determine agricultural products and machine part perimeter measurement. Some systems estimate the length of the boundary around the object by counting the pixels, but some of them use the picture selection method which is based on the selection of the objects manually of images, to investigate the perimeter value [10].

2.2.1.8. *Describing shape*

Shape and size are inseparable in a physical object, and both are necessary if the object is to be satisfactorily described. Further, in defining the shape, some dimensional parameters of the object must be measured. Seeds, grains, fruits, and vegetables are irregular in shape because of the complexity of their specifications, theoretically, that requires an infinite number of measurements [10].

2.2.1.9. *3D measurements*

3D metrics have first been developed for simplification purposes, but three-dimensional image assessments cause new challenges. There is measurement distortion between an original 3D surface and its twisted version. The other significant problem is an analysis of 3D views by using 2D screens. Image measures still require identifying the pixels that are linked to each other [10].

2.2.2. *The experimental methodology*

Ankara University, Faculty of Agriculture, Department of Agricultural Machinery and Technologies students helped for the hand dimension determinations. The specifications of these 25 students were recorded as 171 cm average height, 64 kg average weight, and 23 years average age, respectively. The student's hand measurements had been done by using a caliper with 1 mm precision. The caliper measurement results are compared with video analysis to check the accuracy of the determination, and the difference between the results was analyzed statistically.

2.2.3. *Video processing method of sample application*

The software development process has been done by using LabVIEW graphical program interface tools. For this purpose, LabVIEW Vision module and its sub-modules have been used. The LabVIEW Vision module contains video processing and motion detection codes. At Vision sub-module Vision Acquisition Software (VAS) was used for imaging and video acquisition functions. Also, Vision Development Module (VDM) was used for image processing and analysis functions.

In this application, a template of the selected object was created by using Vision Assistant elements which are seen in the display area. By this way, the template position is determined by using the developed software (**Figures 7** and **8**).

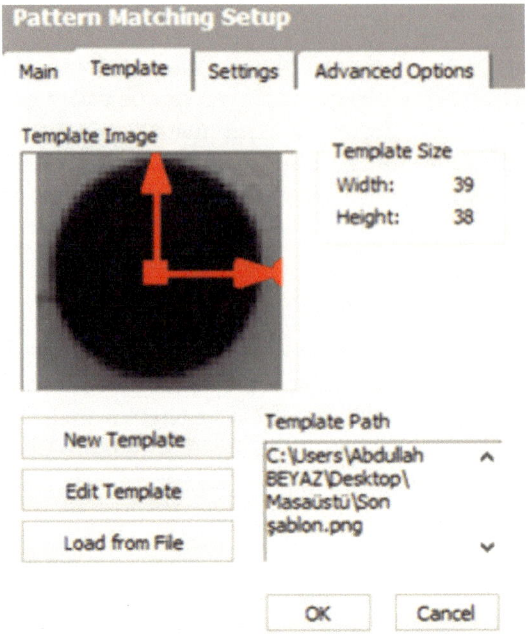

Figure 7. Defined template.

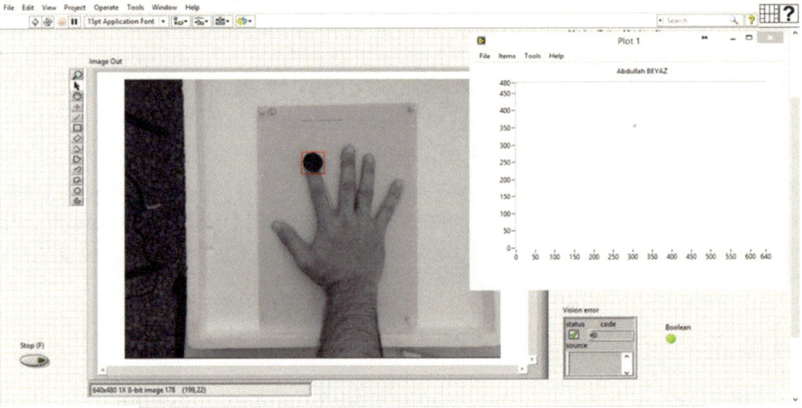

Figure 8. Object and coordinate detection by using LabVIEW software.

Webcam measurement has been done from 50 cm distance. The specified object has been detected automatically at the two-dimensional axis and positioned by using Math script module (**Figure 9**).

Measurement process has been started with measurement of the thumb finger as the first finger and measurement of all other fingers, respectively.

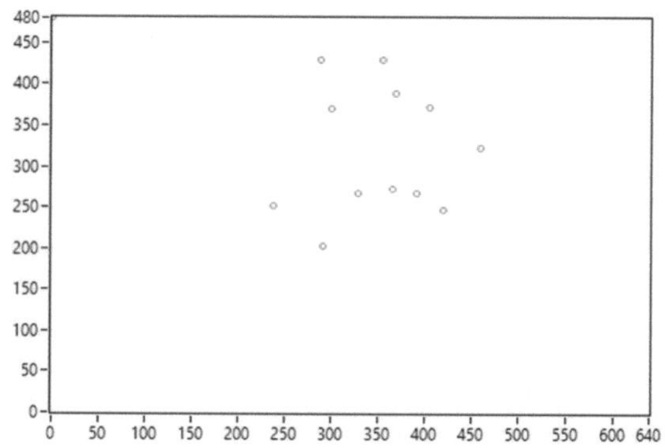

Figure 9. Determined locations on graph.

2.2.3.1. The characteristics of the dataset

All experimental measurements had been performed with a caliper with 1 mm precision. In this sample application, a USB webcam (Logitech C920) is used which has 1920 × 1080 pixel resolution, USB 3.0 interface unit, and video capture capability up to 30 frames per second (**Figure 6**). Webcam resolution is set to 640 × 480 pixels and 30 frames per second for the software working speed competence. Also, computational results obtained from the software under 1 second as (x_1–y_1, x_2–y_2) coordinate data. Coordinate ranges of the measurement graph are set as 640 × 480 for the real resolution comparison for the vision area. Then this (x_1–y_1, x_2–y_2) coordinate data are used for each finger dimension evaluation. Caliper measurements and video coordinate data are analyzed by using regression analysis and also prediction equations added to this regression results for understanding the efficiency of video measurement system.

3. Results and discussion

The studies in the literature support the research findings. The literature studies show that the high success rate can be achieved in this kind of researches. But in this study, parallax errors affected the results. So, in this application, the positions of the fingers affected the webcam parallax errors. These parallax errors can be noticed from the results.

When we look at **Figures 10–14**, it is seen that the value of the regression is 76.9% for the first finger, 88% for the second finger, 68.2% for the third finger, 75.4% for the fourth finger, and 85% for the fifth finger.

Alike the obtained results of this application, Beyaz [8] stated that measurement of regression value rates between 68.2% and 93.2% which is obtained from hand photos.

Epak et al. [11] stated that they developed plate recognition system by using an ordinary computer, camera, LabVIEW 8.2.1 Vision Assistant, and LabVIEW 11.0, and success of the system is 98%.

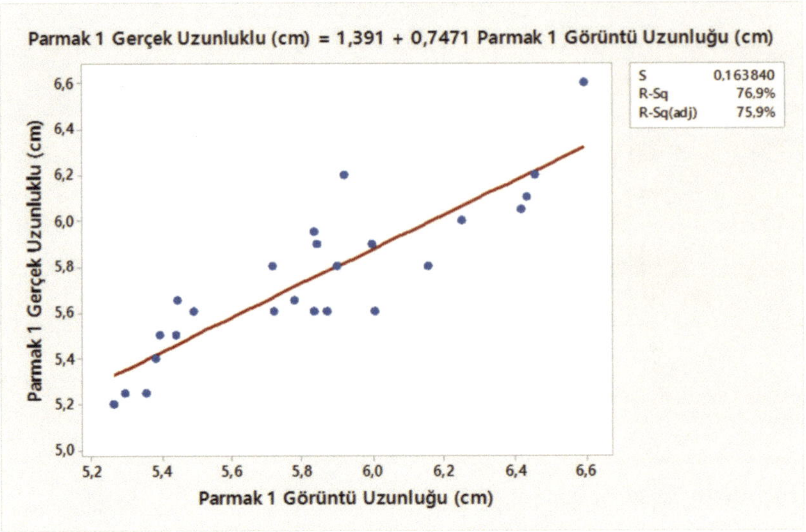

Figure 10. Relationship between thumb finger caliper measurements as the first finger.

Mertens et al. [12] in their research stated that they had used 100 clean and 100 dirty eggs for distinguishing clean ones. They used a camera capable of 30 frames per second with 640 × 480 resolution for the software which they developed in LabVIEW 6.0 platform. They also stated that examples have 91% accuracy.

Velasco et al. [13] studied on shape detection using Kinect; they find the shape detection effectiveness on LabVIEW platform as 100%.

Raut and Ingole [14] worked on detecting leaf diseases using image analysis; they made the plant disease detection accurate with high precision 93%.

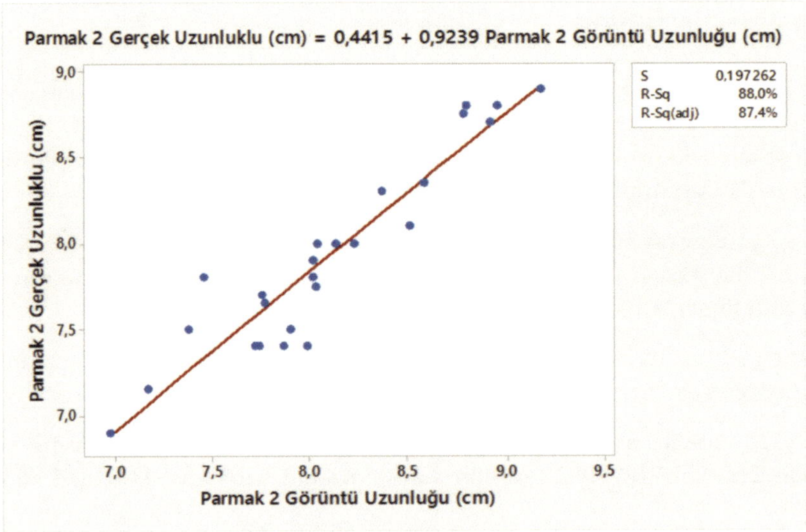

Figure 11. Relationship between second finger caliper measurements.

http://dx.doi.org/10.5772/intechopen.71236

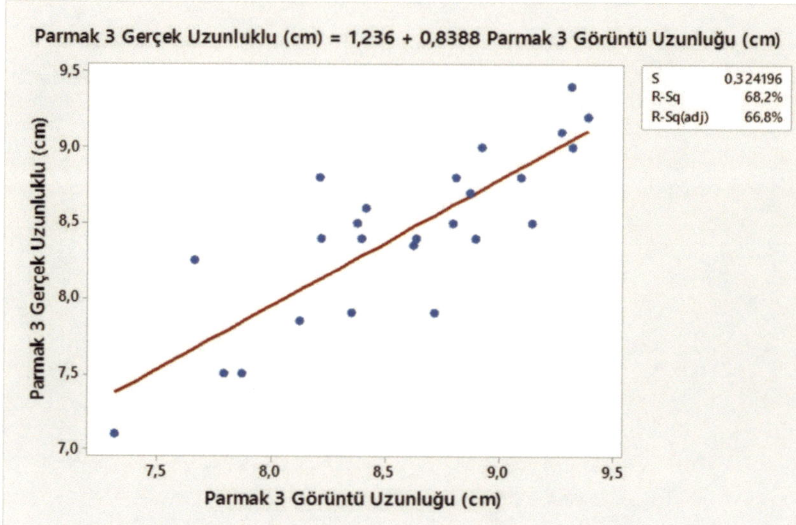

Figure 12. Relationship between third finger caliper measurements.

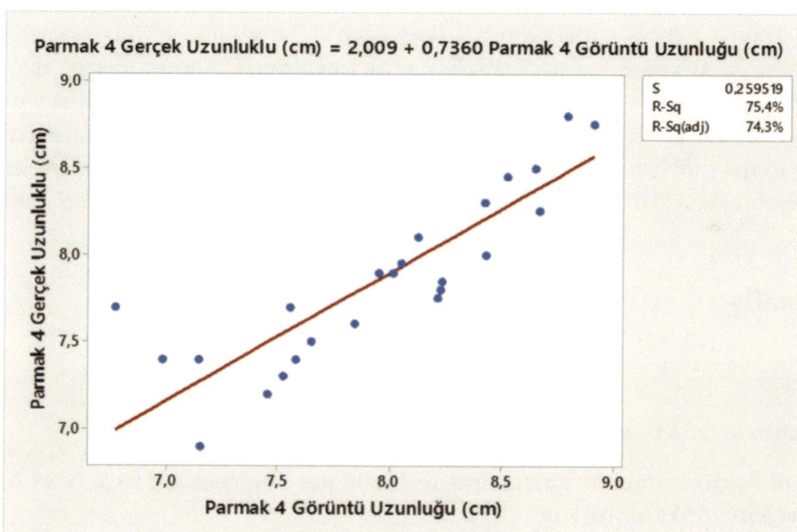

Figure 13. Relationship between fourth finger caliper measurements.

Rob et al. [15] studied on motion tracking by using a web cam on LabVIEW; they stress that they are tracking the defined object effectively.

4. Conclusions

In this chapter, digital video processing technique has been explained. At first, basic requirements about the video processing were discussed for understanding video analysis and acquisition system

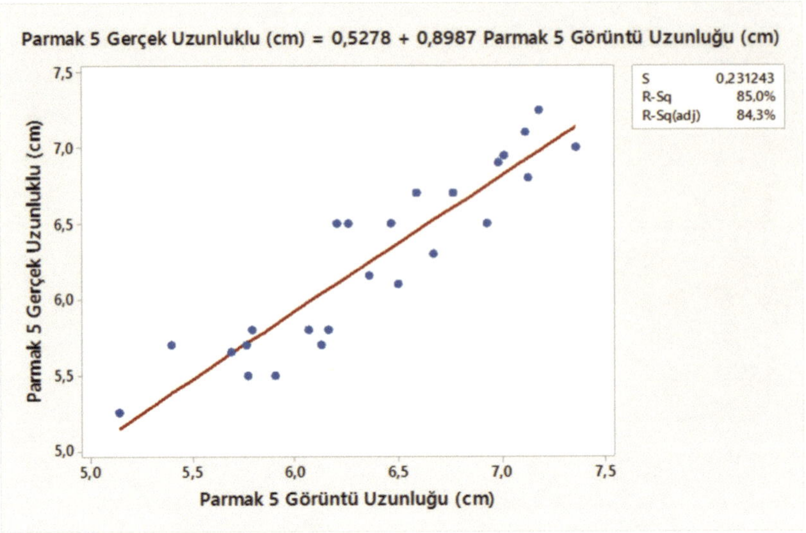

Figure 14. Relationship between fifth finger caliper measurements.

of LabVIEW. Then a sample application has been given. The results of video analysis of sample application have been presented and parallax errors explained. The results of real-time video measurements show that the technique works fast and easy in many fields. When we look at this technology, we need to use computer and electronic-based systems. These systems promise a future as a solution to the problems. By this way, we can develop more efficient machine and tools for agricultural applications like fertilizing, pesticide applications, product sorting, and classification.

Author details

Abdullah Beyaz

Address all correspondence to: abeyaz@ankara.edu.tr

Department of Agricultural Machinery and Technologies Engineering, Faculty of Agriculture, Ankara University, Ankara, Turkey

References

[1] Ün MO. LabVIEW based target tracking system development [master thesis] T. C. Marine War College, Institute of Marine Sciences and Engineering, Department of Electronics Systems Engineering; 2013 Turkish Republic Council of Higher Education Thesis Center

[2] Russ JC. The Image Processing Handbook. fifth ed. CRC Press; 2006. ISBN: 0-8493-7254-2

[3] Anonymous. IMAQ Vision for LabVIEW User Manual. 2015. http://www.ni.com/pdf/manuals/371007a.pdf [Accessed: 29.03.2015]

[4] Szabo R, Gontean A, Lie I. Cheap Live Color Recognition with Webcam. 2011. 978-1-4577-0746-9/11/$26.00 ©2011 IEEE

[5] Pal Singh A, Julka N. Screw classification using machine vision. IJECT. 2011;**2**(1):31-33. ISSN: 2230-7109 (Online)

[6] Ravi Kumar AV, Nataraj KR, Rekha KR. Morphological Real-Time Video Edge Detection in LabVIEW. (IJCSIT) International Journal of Computer Science and Information Technologies. 2012;**3**(2):3808-3811

[7] Dilip R, Bhagirathi V. Image processing techniques for coin classification using LabVIEW. Open Journal of Artificial Intelligence. 2013, 2013;**1**(1):13-17. DOI: 10.12966/ojai.08.03.2013

[8] Beyaz A, Eminoğlu MB, Acar Aİ, Öztürk R, Turker U. Determination of agricultural workers hand anthropometry using image analysis method. 16th National Ergonomics Congress, 3-5 December 2010, Çorum. 2010. Book of abstracts 22 p.

[9] Küçüker A. What is LabVIEW?. 2013. http://lab-view.com/Labviewned/ [Accessed: 31.07.2014]

[10] Beyaz A. Advanced Topics in Measurements. Chapter 7: Machine Vision Measurement Technology and Agricultural Applications. InTech; 2012. ISBN: 978-953-51-0128-4. http://www.intechopen.com/books/advanced-topics-in-measurements

[11] Epak K, Kaushik M, Vashishath M. License plate recognition system based on image processing using LabVIEW. International Journal of Electronics Communication and Computer Technology (IJECCT). 2012;**2**(4):183-188

[12] Mertens K, De Ketelaere B, Kamers B, Bamelis FR, Kemps BJ, Verhoelst EM, De Baerdemaeker JG, Decuypere EM. Dirt detection on brown eggs by means of color computer vision. Poultry Science. 2005;**84**:1653-1659

[13] Velasco N, David Rivas L, Galarza EE. Object detection and comparison of different shapes and materials using Kinect. Journal of Telecommunication Electronic and Computer Engineering. 2017;**9**(1–5):97-100. e-ISSN: 2289-8131

[14] Raut S, Ingole K. Review on leaf disease detection using image processing techniques. International Research Journal of Engineering and Technology (IRJET), 2017;**04**(04):2044-2047. e-ISSN: 2395-056

[15] Rob R, Tirian GO, Panoiu M. LabVIEW application for motion tracking using USB camera. IOP Conference Series: Materials Science and Engineering. 2017;**200**:012008

Chapter 7

Diffusion-Steered Super-Resolution Image Reconstruction

Baraka J. Maiseli

Additional information is available at the end of the chapter

http://dx.doi.org/10.5772/intechopen.71024

Abstract

For decades, super-resolution has been a widely applied technique to improve the spatial resolution of an image without hardware modification. Despite the advantages, super-resolution suffers from ill-posedness, a problem that makes the technique susceptible to multiple solutions. Therefore, scholars have proposed regularization approaches as attempts to address the challenge. The present work introduces a parameterized diffusion-steered regularization framework that integrates total variation (TV) and Perona-Malik (PM) smoothing functionals into the classical super-resolution model. The goal is to establish an automatic interplay between TV and PM regularizers such that only their critical useful properties are extracted to well pose the super-resolution problem, and hence, to generate reliable and appreciable results. Extensive analysis of the proposed resolution-enhancement model shows that it can respond well on different image regions. Experimental results provide further evidence that the proposed model outperforms.

Keywords: super-resolution, resolution, enhancement, regularization, diffusion

1. Introduction

Before deepening into the super-resolution imaging, let us discuss the term resolution. Most people, particularly those not in the imaging field, define resolution broadly as the physical size of an image. For a two-dimensional digital image, this definition implies an area in the image given as the product of the number of pixels in the horizontal and vertical dimensions (pixel or picture element is the smallest unit of information in a digital image). In this context, therefore, a high-resolution image contains a higher pixel count than a low-resolution image. **Figure 1(a)** includes features with higher perceptual qualities than those in **Figure 1(b)**, but both images have equal sizes. From the figure, therefore, we see that dimension only seems inadequate to define the resolution of an image.

(a) High quality image (b) Low quality (blurred) image

Figure 1. Images of dimensions 2179 × 2011.

Resolution, more generally, means the quality of a scene (image or video). Five major types of image resolutions are known: pixel resolution, spectral resolution, temporal resolution, radiometric resolution, and spatial resolution. The use of these variations depends on the application. Pixel resolution refers to the total number of pixels a digital image contains. Hence, both images in **Figure 1(a)** and **(b)** possess equal pixel resolutions of 2179 × 2011. In other words, each image is approximately 4.4 megapixels (2179 × 2011 = 4,381,969 pixels ≈4.4 megapixels). Unfortunately, pixel count offers fraction of the pieces of information contained in the image. For a colored image with red, green, and blue channels, an individual pixel can only accommodate the details of a single color. Spectral resolution describes the ability of an imaging device to distinguish the frequency (or wavelength) components of an electromagnetic spectrum. Imagine spectral resolution as the degree in which you can uniquely discern two different colors or light sources. Temporal resolution refers to the rate at which an imaging device revisits the same location to acquire data. When dealing with videos, for example, the term implies an average time between consecutive video frames: a standard video camera can record 30 frames per second, implying that every 33 ms, this camera captures an image. In remote sensing, temporal time is usually measured in days to represent time that a satellite sensor revisits a specific location to collect data. Radiometric resolution defines the degree at which an imaging system can represent or distinguish intensity variations on the sensor. Expressed in number of bits (or number of levels), radiometric resolution provides the actual content of information in the image. Spatial resolution explains how an imaging modality can distinguish two objects. In practical situations, spatial resolution describes clarity of an image and defines the resolving power of an image-capturing device. The perceptual quality of an image increases with the spatial resolution. This research presents super-resolution imaging as one of the available techniques to enhance the spatial resolution of an image.

Most people are naturally inclined to high-quality and visually appealing images that contain adequate details. However, this demand is not always achieved because of some imperfections in the imaging process. Therefore, scholars have proposed hardware and software approaches to address the challenge. The former approach requires sensor modification, and it may be achieved by reducing the physical sizes of the pixels—a process that increases pixel density (number of pixels per unit area) on the surface of the sensor [1]. The hardware approach gives perfect resolution enhancement, but the technique endures several drawbacks: (1) it introduces shot noise into the captured images, (2) it makes the imaging device costly and unnecessarily bulkier, and (3) it lowers the charge transfer rate because of the increased chip size [2]. These challenges have prompted scholars to search for software techniques, which are cost-effective and reliable, to improve the spatial resolution of an image without effecting circuitry of the imaging device. In this case, an image can be captured by a low-cost device and processed to generate its corresponding high-quality version.

The classical software approach that has gained a considerable attention of scholars is called super-resolution [3–6], which uses signal processing principles to restore high-resolution images from at least one low-resolution image. Super-resolution techniques can be put into two major categories: single-frame-based, which generates a high-resolution image from the respective single low-resolution image [7, 8], and multi-frame-based, which exploits information from a sequence of degraded images to generate a high-quality image [2, 6]. The current work builds on the multi-frame super-resolution framework, which implicitly encourages noise reduction from the input low-resolution images. The framework bridges total variation (TV) [9] and Perona and Malik [10] smoothing functionals and allows for these functionals to interact in such a way that super-resolution and preservation of critical image features are simultaneously conducted.

2. Image degradation model

The multi-frame super-resolution framework can better be understood through a conceptual degradation model, which shows how an unknown high-resolution image, u, undergoes a variety of degradations to form M low-quality images, y_k, with $k=1,\dots,M$ denoting positions of the low-resolution frames (**Figure 2**). In practice, the degradation process of u to generate y_k involves warping, blurring, decimation (downsampling), and noising, respectively defined in this work by the operators W_k, B_k, D_k, and η_k: warping introduces rotations and translations into u, hence changing its geometrical properties; blurring reduces sharpness of features in u; decimation samples u and lowers its physical size; and noising corrupts u with noise, assumed to be additive.

Figure 2 can be transformed into

$$y_k = W_k B_k D_k u + \eta_k, \tag{1}$$

which explains how the degradation model generates frame k in a set of low-resolution images. The goal of the present study is to estimate u under the degradation conditions, and one approach to achieve the goal is to re-define Eq. (1) into the minimization problem that aims to lower η_k. Therefore, using the L_p norm, where $p\in[1,2]$ (the range $0\leq p<1$ is excluded because the values of p contained in this interval lead to nonconvex minimization problems that are susceptible to unstable solutions), the formulation to optimize u becomes

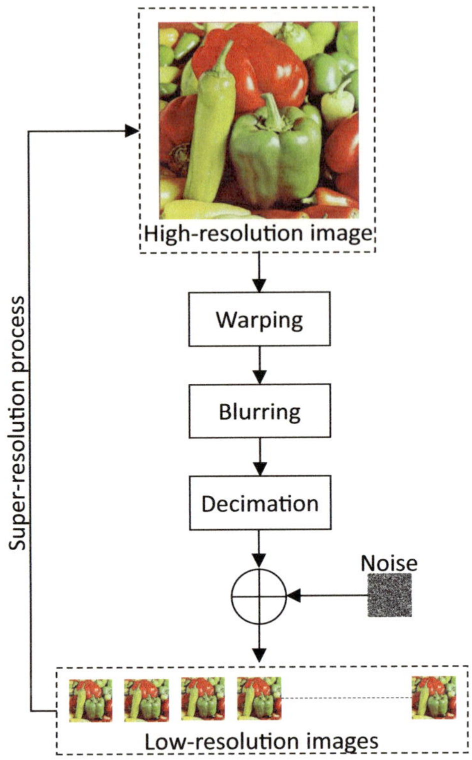

Figure 2. Image degradation model.

$$\min u\left\{E(u) = \frac{1}{2M}\sum_{k=1}^{M} \|W_k B_k D_k u - y_k\|_p^p\right\}, \quad (2)$$

where E is modeled as an energy functional that defines noise level in the degraded image. The gradient of the cost of E in Eq. (2) is

$$J_p = \frac{\partial}{\partial u}\left[\frac{1}{2M}\sum_{k=1}^{M} \|W_k B_k D_k u - y_k\|_p^p\right] \quad (3)$$

$$= \frac{1}{2M}\sum_{k=1}^{M} D_k^T B_k^T W_k^T \mathrm{sign}\left(W_k B_k D_k u - y_k\right) \odot \left|W_k B_k D_k u - y_k\right|^{p-1}, \quad (4)$$

where D_k^T is the upsampling operator, B_k^T and W_k^T are the inverse operators for blurring and warping, respectively, and $\odot$ denotes the Hadamard (element-wise) operator for two matrices. The solution of Eq. (2) can be obtained when $J_p = 0$.

For $p = 1$, Eq. (4) evaluates to

$$J_1 = \frac{1}{2M} \sum_{k=1}^{M} D_k^T B_k^T W_k^T \text{sign}\left(W_k B_k D_k u - y_k\right) = 0, \tag{5}$$

which shows that, after shifting and zero filling, $D_k^T B_k^T W_k^T$ copies values from the low-resolution to the high-resolution images, and $W_k B_k D_k$ reverses the operation [11]. Pixel values are unaffected by these complimentary operations, implying that each entry in J_1 is impacted by entries from all low-resolution images. **Figure 3** shows the influences of D and that of D^T on the reconstructing image. In their work, Farsiu et al. noted that the L_1 minimization in Eq. (5) corresponds to the pixel-wise median, a robust estimator that addresses favorably noise and outliers in the input data. But the L_1 norm is nondifferentiable at zero, a property that makes the minimization process unstable and that generates undesirable solutions.

For $p = 2$, Eq. (4) becomes a solution of the L_2 norm minimization, or

$$J_2 = \frac{1}{2M} \sum_{k=1}^{M} D_k^T B_k^T W_k^T \left(W_k B_k D_k u - y_k\right) = 0, \tag{6}$$

which was proved in [12] that it represents pixel-wise mean of measurements. The L_2 norm is less-robust against erroneous data, but the metric has better mathematical properties: convexity, differentiability, and stability. Therefore, several scholars prefer the L_2 objective functions in situations where data contain low noise as in our case.

The super-resolution problem, whether formulated through L_1 or L_2 norm, has an ill-posedness nature. Given that r is the resolution factor, then for the under-determined case, or for $M < r^2$, and for the square case, or for $M = r^2$, the problem may evaluate to infinitely many undesirable solutions. Also, for the small amount of noise in the data, ill-posed problems tend to introduce larger perturbations in the final solutions. These issues can be effectively addressed through a technique called regularization, which has another advantage of speeding the convergence rate of the evolving solution. This work addresses the super-resolution ill-posedness through regularization functionals from nonlinear diffusion processes, which have been reported that they can preserve important image features (edges, contours, and lines) [13–15]. The proposed regularizer integrates total variation (TV) [9] and Perona and Malik (PM) [10] models that complement one another to generate appealing results.

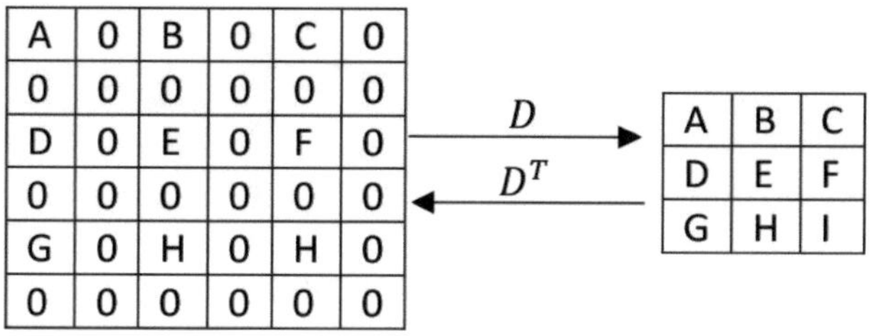

Figure 3. Downsampling matrix, D, and upsampling matrix, D^T, applied on an image. The resolution reconstruction factor used is two for both horizontal and vertical dimensions of the image.

3. Hybrid super-resolution model

3.1. Regularization functionals

Considering the super-resolution ill-posedness property, a hybrid framework combining TV and PM regularization kernels has been formulated. The framework includes additional parameters, α and β, which establish a proper balance between TV and PM during regularization. The objective is to de-emphasize weaknesses of the models and amplify their strengths so that the super-resolved images are superior.

In [9], Rudin et al. established the TV model that explains how noise in the image can be reduced. The model is based on the fact that a noisy image contains a higher total variation, defined by the integral of the absolute gradient of the image or

$$\rho(|\nabla u|) = \int_{\Omega} |\nabla u| dx, \tag{7}$$

where ρ is the TV energy functional, Ω defines the domain under which u exists, and x denotes the two-dimensional spatial coordinate on Ω. Therefore, reducing noise is equivalent to minimizing ρ. Being defined in the bounded variation space, TV functionals allow for discontinuities in the image functions. Hence, regularization through TV promotes recovery of edges, which appear as "jumps" or discontinuous parts of the image, and effective noise removal. But studies have revealed that TV formulations favor piecewise-constant solutions, a consequence that generates staircase effects and introduces false edges [16]. Also, TV regularization tends to lower contrast even in noise-free or flat image regions [17].

In the similar notion of the TV principle, Perona and Malik proposed an energy functional, φ, defined by

$$\varphi(|\nabla u|) = \frac{K^2}{2} \int_{\Omega} log\left(1 + \left(\frac{|\nabla \mathrm{u}|}{K}\right)^2\right) dx, \tag{8}$$

where K denotes the shape-defining constant, which can be minimized to suppress noise [10]. Minimizing Eq. (8), which originates from robust statistics, produces a nonlinear diffusion equation that embeds a fractional conduction coefficient for preserving edges. The PM energy functional in Eq. (8) is nonconvex for $|\nabla u| > K$, an undesirable property that can generate instabilities in the evolving solution. This work presents a technique that retains the convex portion, $|\nabla u| \leq K$, and complements the nonconvex portion of the PM potential by the TV energy functional.

The regularization process is often supported by the fidelity potentials

$$\psi(u) = \frac{\lambda}{2} \int_{\Omega} (u - f)^2 dx \tag{9}$$

for additive noise, $f = u + \eta$, and

$$\phi(u) = \lambda \int_{\Omega} \left(\log u + \frac{f}{u} \right) dx \tag{10}$$

for multiplicative noise [18], $f = u\eta$, where f is the corrupted image and λ is the fidelity parameter that balances the trade-off between u and f. The fidelity term is often added to the regularization framework.

3.2. Proposed super-resolution model

The hybrid model can be derived from the minimization problem that integrates the corresponding energy functionals from super-resolution, TV, PM, and fidelity. Assuming additive noise and L_2 estimator for the super-resolution part, the (regularized) minimization super-resolution problem parametrized in α and β becomes

$$\min u \left\{ H(u, |\nabla u|) = \frac{1}{2M} \sum_{k=1}^{M} \| W_k B_k D_k u - y_k \|_2^2 + \alpha \rho(|\nabla u|) + \beta \varphi(|\nabla u|) + \psi(u) \right\}, \tag{11}$$

where $\alpha, \beta \in [0, 1]$ and $\beta = \overline{\alpha}$. Solving Eq. (11) using the Euler-Lagrange equation, and embedding the result into the time-dependent system gives

$$\frac{\partial u}{\partial t} = \frac{1}{2M} \sum_{k=1}^{M} D_k^T B_k^T W_k^T (W_k B_k D_k u - y_k) + \operatorname{div}\left(\frac{\alpha}{|\nabla u|} \nabla u \right) + \operatorname{div}\left(\frac{\beta}{1 + \left(\frac{|\nabla u|}{K} \right)^2} \nabla u \right) - \lambda (u - f). \tag{12}$$

Eq. (12) offers both super-resolution image reconstruction and noise removal capabilities, dictated by TV and PM models. From the equation, as $t \to \infty$, u approaches an optimal solution—a stationary function that solves the energy functional, H, in Eq. (11). Eq. (12) has interesting properties for various parts of the image: in flat regions ($|\nabla u| \to 0$), Eq. (12) reduces to

$$\frac{\partial u}{\partial t} = \frac{1}{2M} \sum_{k=1}^{M} D_k^T B_k^T W_k^T (W_k B_k D_k u - y_k) + (\alpha C + \beta) \Delta u - \lambda (u - f), \tag{13}$$

where $C > 0$ is a constant. This equation has a Laplacian term, Δu, which possesses isotropic diffusion characteristics to strongly and uniformly suppress noise in flat regions. In the neighborhood of the edges ($|\nabla u| \to \infty$), Eq. (12) becomes

$$\frac{\partial u}{\partial t} = \frac{1}{2M} \sum_{k=1}^{M} D_k^T B_k^T W_k^T (W_k B_k D_k u - y_k) - \lambda (u - f), \tag{14}$$

implying protection of edges against smoothing. This automatic interplay between reconstruction and regularization components helps to generate superior super-resolved images.

3.3. Numerical implementation

The solution of the proposed super-resolution model in Eq. (12) was iteratively estimated using the steepest descent method. Therefore, the evolution equation in Eq. (12) can be converted into a numerical system

$$u_{n+1} = u_n - \tau \left\{ \frac{1}{2M} \sum_{k=1}^{M} D_k^T B_k^T W_k^T \left(W_k B_k D_k u_n - y_k \right) + \operatorname{div}\left(\frac{\alpha}{|\nabla u_n|} \nabla u_n \right) + \operatorname{div}\left(\frac{\beta}{1 + \left(\frac{|\nabla u_n|}{K} \right)^2} \nabla u_n \right) - \lambda \left(u_n - f_n \right) \right\}, \quad (15)$$

where n denotes the iteration number that defines the solution space index of u, and $\tau > 0$ denotes constant of the step size in the gradient direction. To encourage stability of the evolution equation in (15), the Courant-Friedrichs-Lewy condition, that is $0 < \tau \leq 0.25$, should be satisfied [19]. From the equation, the degradation matrices, namely W_k, B_k,and D_k, and their corresponding transpose versions may be regarded as direct operators for image manipulations: shifting, blurring, and downsampling, along with the reverse of these operations [11]. With this observation of the matrices properties, implementation of the super-resolution component of Eq. (15) can be achieved using cascaded operators without explicitly constructing the

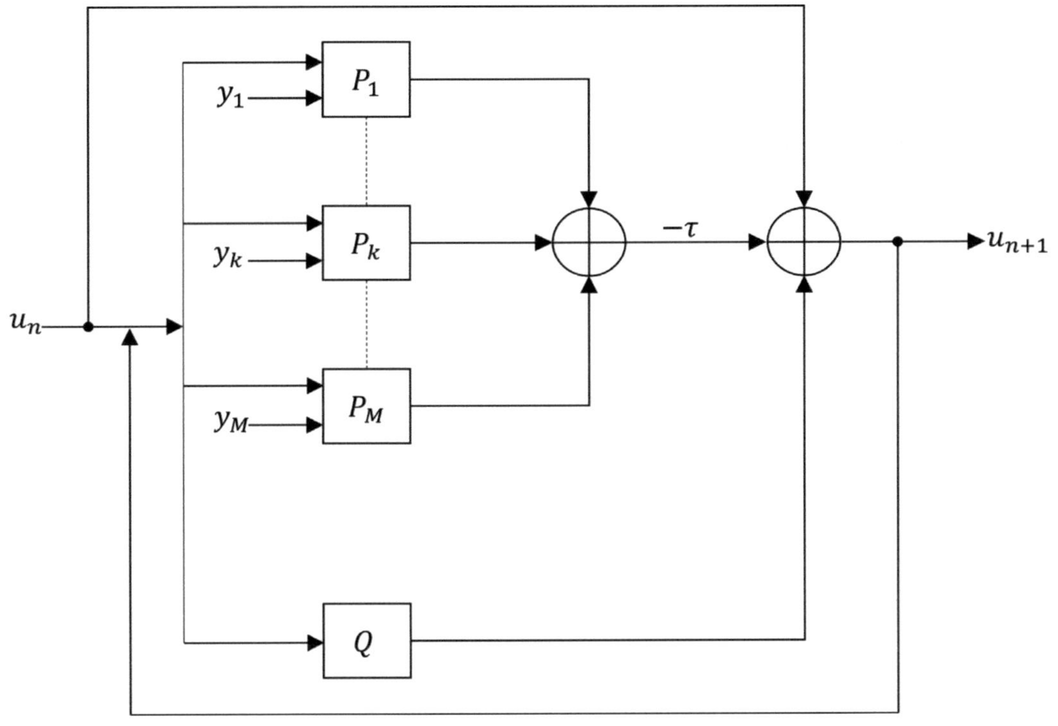

Figure 4. Block diagram representation of the proposed super-resolution model. The blocks P_k and Q are defined in **Figures 5** and **6**.

operators as matrices. This implementation strategy helps to boost the algorithmic speed and to optimize hardware resources.

Eq. (15) can be represented in block form by **Figure 4**. From the Figure, each low-resolution frame, y_k, is compared with the current estimate, u_n, of the high-resolution image. This process is undertaken by block P_k, detailed in **Figure 5**—an operator that represents the gradient back projection to compare the kth degraded frame and the high-resolution estimate at the nth

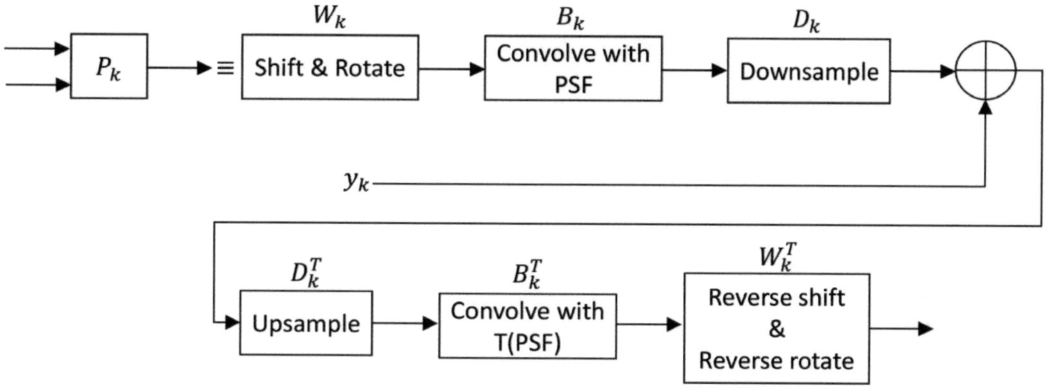

Figure 5. Extended block diagram representation of the similarity cost derivative, P_k, in **Figure 4**.

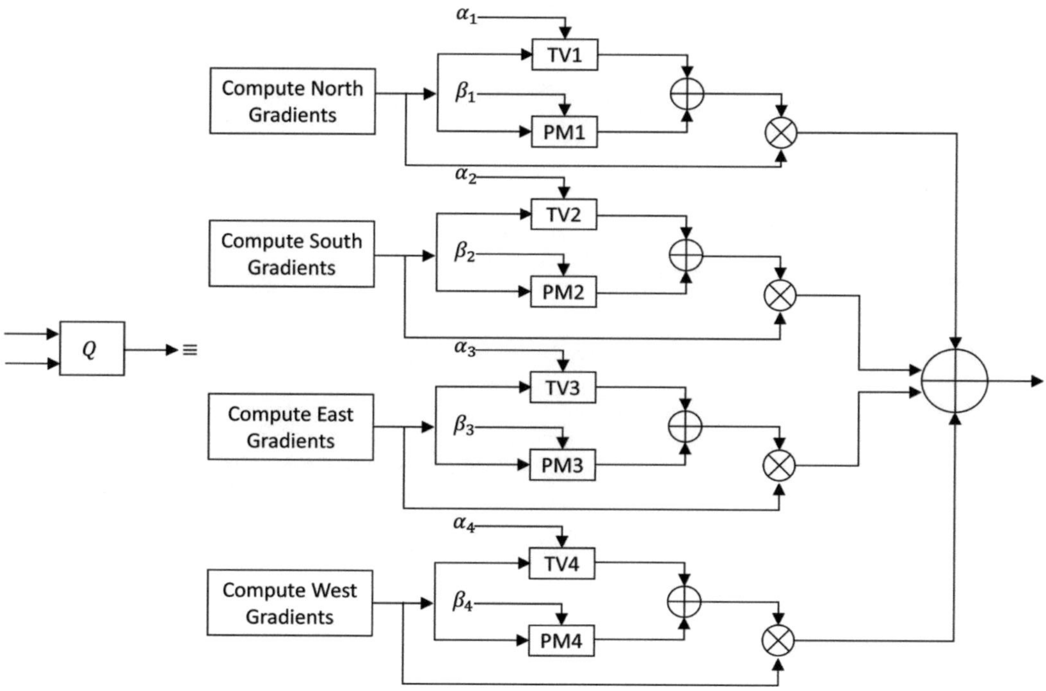

Figure 6. Block diagram representation of the smoothing cost derivative, Q, in **Figure 4**.

iteration of the steepest descent method. Note from **Figure 5** that T(PSF), with PSF denoting the point spread function, replaces B_k^T with a simple convolution operator. This block can be implemented by flipping, on the respective axes, rows, and columns of the PSF in the up-down and left-right directions, respectively. Gradient of the regularization term is represented by block Q, defined more explicitly in **Figure 6**, which ensures that the evolution process converges and gives desirable solutions.

4. Experimental methodology

Several experiments were executed to determine performance of the proposed super-resolution model relative to the classical approaches. The methodology and procedures under which the experiments were undertaken can be explained as follows: firstly, high-resolution images of bike, butterfly, flower, hat, parrot, Parthenon, plant, and raccoon (**Figure 7**) were degraded to generate the corresponding low-resolution images (**Figure 8**, first column). Note that the original images were downloaded from the public domain with standard test images.[1] These images were selected because they contain detailed features, and hence it would be easier to test the superiority of various super-resolution methods. As an example, the "Raccoon" image contains small-scale features (fine textures or fur) that most super-resolution approaches may find hard to restore. Degradation of the original images was achieved through warping, blurring, decimation, and noise addition to create sequences of 10 low-quality images with consecutive pairs differing by some rotation and translation motions. To void impacts of registration errors on the reconstruction process, the

(a) Bike (b) Butterfly (c) Flower (d) Hat

(e) Parrot (f) Parthenon (g) Plant (h) Raccoon

Figure 7. Original high-resolution images.

[1]http://www4.comp.polyu.edu.hk/~cslzhang/NCSR.htm

Figure 8. Super-resolution results from different methods.

warping matrix was fixed. Thus, for 10 multiple low-resolution images, the warping matrix for the horizontal and vertical displacements, respectively denoted by Δx and Δy, was defined as follows:

Δx	0.56	1.03	0.85	0.32	−0.45	−0.43	0.92	1.23	0.93	0.64
Δy	0.12	0.53	0.27	0.00	−0.83	1.12	1.08	0.12	0.54	1.37

Next, super-resolution methods based on a variety of regularizers, namely NC00 [20], TV [9], ANDIFF [21], and Hybrid, were applied on the degraded images to restore their original versions. Lastly, the objective metric, namely feature similarity (FSIM) [22], and the subjective metric were used to compare performances of different methods. FSIM incorporates into its formulation some aspects of the human visual system, and hence the metric is considered superior over several other existing image quality metrics. A visually appealing image has a higher value of FSIM, and vice versa.

5. Results and discussions

Visual results show that the classical methods tend to add undesirable artificial features into the reconstructed images (**Figure 8**). For instance, NC00 introduces bubble-like features around borders, edges, and corners, which are the critical features that emulate the human visual system. The method, on the other hand, does well on homogeneous image regions. The super-resolution method based on TV produces relatively sharper images, but the method also adds artifacts on homogeneous parts of the final images—an effect that degrades the visual quality of the images. The ANDIFF method generates smoother results that contain little artifacts, but the method underperforms for highly-textured images such as the Raccoon. The proposed hybrid model established a proper balance between smoothness and critical feature preservation (**Figure 8**, last column). Visually, the reconstructed images by our approach are more natural and are free from obvious artifacts. One may argue about a slight blurriness in our results. However, given the higher capability of the proposed method to preserve sensitive image features, this effect may be ignored. Also, the line graphs (taken near the last row across all columns) further confirm that the proposed method is superior because it generates a one-dimensional curve that closely matches the original one (**Figure 9**).

Numerical results demonstrate that, in all cases of the input images, the proposed super-resolution method achieves higher quality values (**Table 1**). These convincing objective observations can be explained well from the new formulation in Eq. (12): the hybrid super-resolution model captures the qualities of both PM and TV, an advantage that may promote higher objective quality results. Besides, our formulation incorporates parameters that give an effective interplay between the regularization functionals.

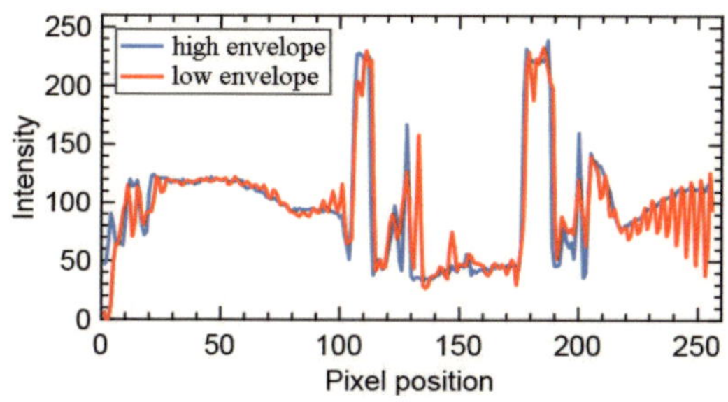

(a) NC00

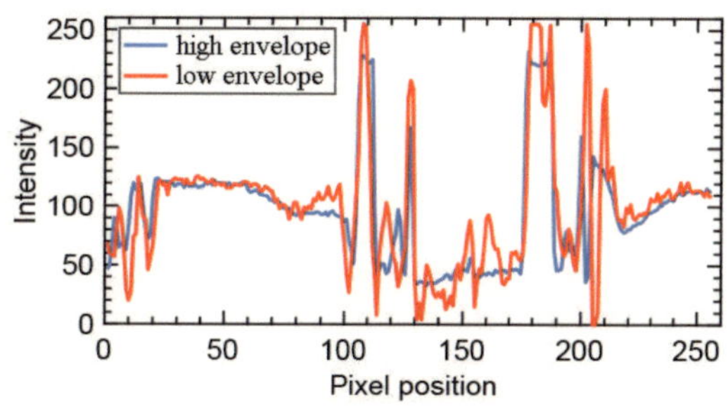

(b) TV

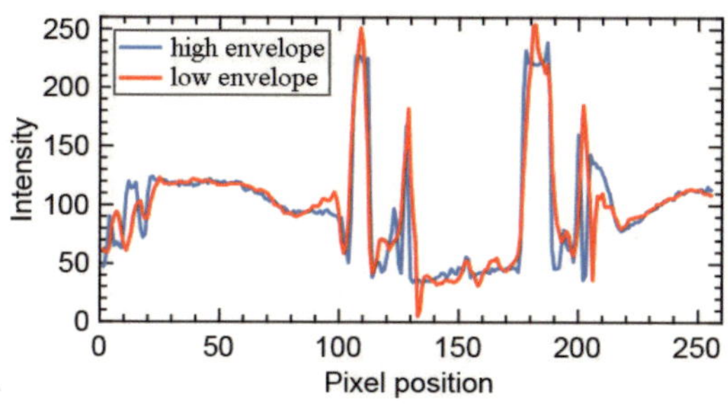

(c) ANDIFF

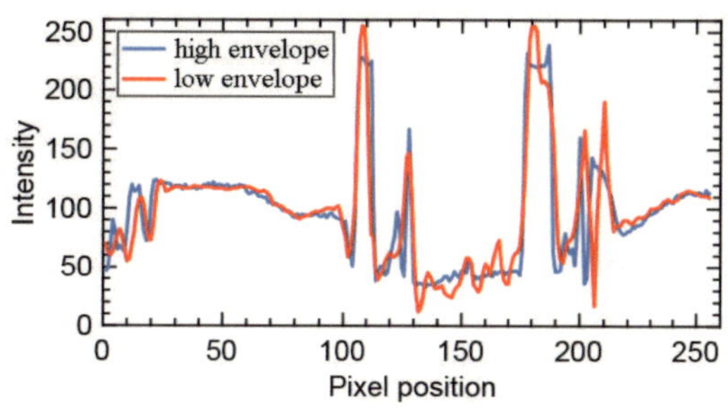

(d) Our method

Figure 9. Line graphs of images generated by different super-resolution methods.

Image	NC00	TV	ANDIFF	Proposed method
Bike	0.7139	0.7148	0.7386	0.7642
Butterfly	0.6721	0.6733	0.7386	0.7592
Flower	0.6998	0.7084	0.7669	0.7970
Hat	0.7512	0.7624	0.8106	0.8194
Parrot	0.7672	0.7908	0.8595	0.8738
Parthenon	0.7101	0.7287	0.7450	0.7618
Plant	0.7429	0.7416	0.8230	0.8401
Raccoon	0.7591	0.7877	0.8046	0.8257

Table 1. Feature similarities of images restored from various super-resolution methods.

6. Conclusion

In this work, we have established a hybrid super-resolution framework that combines desirable features of TV and PM models. The framework has been parametrized to mask weaknesses of the models, introduce an automatic interplay between TV and PM regularizations, and promote appealing results. More emphasis was put on super-resolving low-quality images while retaining their naturalness and preserving their sensitive image features. Experimental results demonstrate that the proposed framework generates superior objective and subjective results.

Author details

Baraka J. Maiseli

Address all correspondence to: barakamaiseli@yahoo.com

Department of Electronics and Telecommunication Engineering, College of Information and Communication Technologies, University of Dar es Salaam, Dar es Salaam, Tanzania

References

[1] Park SC, Park MK, Kang MG. Super-resolution image reconstruction: A technical overview. IEEE Signal Processing Magazine. 2003;**20**(3):21-36

[2] Maiseli BJ, Elisha OA, Gao H. A multi-frame super-resolution method based on the variable-exponent nonlinear diffusion regularizer. EURASIP Journal on Image and Video Processing. 2015;**2015**(1):22

[3] Dong C, et al. Image super-resolution using deep convolutional networks. IEEE Transactions on Pattern Analysis and Machine Intelligence. 2016;**38**(2):295-307

[4] El Mourabit I, et al. A new denoising model for multi-frame super-resolution image reconstruction. Signal Processing. 2017;**132**:51-65

[5] Peleg T, Elad M. A statistical prediction model based on sparse representations for single image super-resolution. IEEE Transactions on Image Processing. 2014;**23**(6):2569-2582

[6] Zeng X, Yang L. A robust multiframe super-resolution algorithm based on half-quadratic estimation with modified BTV regularization. Digital Signal Processing. 2013;**23**(1):98-109

[7] Purkait P, Pal NR, Chanda B. A fuzzy-rule-based approach for single frame super resolution. IEEE Transactions on Image Processing. 2014;**23**(5):2277-2290

[8] Yang M-C, Wang Y-CF. A self-learning approach to single image super-resolution. IEEE Transactions on Multimedia. 2013;**15**(3):498-508

[9] Rudin LI, Osher S, Fatemi E. Nonlinear total variation based noise removal algorithms. Physica D: Nonlinear Phenomena. 1992;**60**(1–4):259-268

[10] Perona P, Malik J. Scale-space and edge detection using anisotropic diffusion. IEEE Transactions on Pattern Analysis and Machine Intelligence. 1990;**12**(7):629-639

[11] Farsiu S, et al. Fast and robust multiframe super resolution. IEEE Transactions on Image Processing. 2004;**13**(10):1327-1344

[12] Elad M, Hel-Or Y. A fast super-resolution reconstruction algorithm for pure translational motion and common space-invariant blur. IEEE Transactions on Image Processing. 2001; **10**(8):1187-1193

[13] Ma H, Nie Y. An edge fusion scheme for image denoising based on anisotropic diffusion models. Journal of Visual Communication and Image Representation. 2016;**40**:406-417

[14] Tsiotsios C, Petrou M. On the choice of the parameters for anisotropic diffusion in image processing. Pattern Recognition. 2013;**46**(5):1369-1381

[15] Xu J, et al. An improved anisotropic diffusion filter with semi-adaptive threshold for edge preservation. Signal Processing. 2016;**119**:80-91

[16] Chen Y, Levine S, Rao M. Variable exponent, linear growth functionals in image restoration. SIAM Journal on Applied Mathematics. 2006;**66**(4):1383-1406

[17] Chan TF, Esedoglu S. Aspects of total variation regularized L 1 function approximation. SIAM Journal on Applied Mathematics. 2005;**65**(5):1817-1837

[18] Lv X-G, et al. A fast high-order total variation minimization method for multiplicative noise removal. Mathematical Problems in Engineering. 2013;**2013**

[19] Courant R, Friedrichs K, Lewy H. On the partial difference equations of mathematical physics. IBM Journal of Research and Development. 1967;**11**(2):215-234

[20] Pham TQ, Van Vliet LJ, Schutte K. Robust fusion of irregularly sampled data using adaptive normalized convolution. EURASIP Journal on Advances in Signal Processing. 2006;**2006**(1): 083268

[21] Maiseli BJ, Ally N, Gao H. A noise-suppressing and edge-preserving multiframe super-resolution image reconstruction method. Signal Processing: Image Communication. 2015; **34**:1-13

[22] Zhang L, et al. FSIM: A feature similarity index for image quality assessment. IEEE Transactions on Image Processing. 2011;**20**(8):2378-2386

Chapter 8

A New Pansharpening Approach for Hyperspectral Images

Chiman Kwan, Jin Zhou and Bence Budavari

Additional information is available at the end of the chapter

http://dx.doi.org/10.5772/intechopen.71023

Abstract

We first briefly review recent papers for pansharpening of hyperspectral (HS) images. We then present a recent pansharpening approach called hybrid color mapping (HCM). A few variants of HCM are then summarized. Using two hyperspectral images, we illustrate the advantages of HCM by comparing HCM with 10 state-of-the-art algorithms.

Keywords: hyperspectral images, pansharpening, hybrid color mapping, sparsity, image fusion

1. Introduction

Hyperspectral (HS) images have found a wide range of applications in terrestrial and planetary missions. NASA is planning a HyspIRI mission [1–4] that will perform vegetation monitoring for the whole Earth. The spatial resolution of HyspIRI is 60 m. The Compact Reconnaissance Imaging Spectrometer for Mars (CRISM) [5] hyperspectral (HS) imager with 100-m resolution has been monitoring Mars surface since 2006. Although the above imagers have resolution good enough for their respective missions, many other applications such as drought monitoring, fire damage assessment, etc., require higher resolutions. Other notable applications of multispectral and hyperspectral images include target detection [6–13], anomaly and change detection [14–23], tunnel monitoring [24, 25], and Mars exploration [26, 27].

Pansharpening of hyperspectral images usually refers to the fusion of a high-resolution (HR) panchromatic (pan) band with a low-resolution (LR) hyperspectral image cube. A generalization of the above is the fusion of high-resolution multispectral bands with low-resolution hyperspectral bands. According to Loncan et al. [28], pansharpening techniques for HS images can be classified into the following categories. The first category is the component substitution (CS) approach. Well-known CS approaches include Principal Component Analysis (PCA) [29],

Gram-Schmidt (GS) [30], GS Adaptive (GSA) [31], and others. The CS approach is based on the substitution of a component with the pan image. The second category is the multiresolution analysis (MRA) approach. The MRA approach relies on the injection of spatial details that are obtained through a multiresolution decomposition of the pan image into the resampled hyperspectral bands. Some well-known algorithms in this category are Modulation Transfer Function Generalized Laplacian Pyramid (MTF-GLP) [32], MTF-GLP with High-Pass Modulation (MTF-GLP-HPM) [33], Hysure [34, 35], and Smoothing Filter-based Intensity Modulation (SFIM) [36]. The third category contains the hybrid approaches that use concepts from different classes of methods, namely from CS and MRA ones. For example, guided filter PCA (GFPCA) [37] belongs to this group. The fourth category involves Bayesian inference, which interprets the fusion process through the posterior distribution of the Bayesian fusion model. Due to the ill-posed fusion problem, the Bayesian methodology can easily regularize the problem by defining a prior distribution for the scene of interest. Exemplar algorithms include Bayesian naive [38] and Bayesian sparse [39]. The fifth category is known as non-negative matrix Factorization (NMF). Representative methods in this category include the coupled non-negative matrix factorization (CNMF) [40] method.

In addition to the above five categories, we noticed that deep learning-based approaches have been investigated in recent years to address pansharpening for hyperspectral images. In Licciardi et al. [41], the authors proposed the first deep learning-based fusion method, which used autoencoder. In 2015, a modified sparse tied-weights denoising autoencoder was proposed by Huang et al. [42]. The authors assumed that there exists a mapping function between LR and HR images for both pan and MS or HS. During the training process, an LR pan was generated by the interpolated MS, then the mapping function was learned by the given LR pan patches as the input and HR pan patches as the output. In 2016, a supervised three-layer convolutional neural network was proposed by [43] to learn the mapping function between the input HR pan with the interpolated LR MS, and the output HR MS. In Qu et al. [44], a new pansharpening algorithm based on deep learning autoencoder was proposed. Preliminary pansharpening results using both hyperspectral and multispectral images are encouraging.

From the application viewpoint, we categorize the various pansharpening algorithms into three groups. Group 1 methods include coupled non-negative matrix factorization (CNMF) [40], Bayesian naive [38], and Bayesian sparse [39]. These methods require the point spread function (PSF) to be available and they perform better than other methods in most cases. Group 2 methods do not require PSF and contain Principal Component Analysis (PCA) [29], guided filter PCA (GFPCA) [37], Gram-Schmidt (GS) [30], GS Adaptive (GSA) [31], Modulation Transfer Function Generalized Laplacian Pyramid (MTF-GLP) [32], MTF-GLP with High-Pass Modulation (MTF-GLP-HPM) [33], Hysure [34, 35], and Smoothing Filter-based Intensity Modulation (SFIM) [36]. Group 3 methods use only the LR HS images and contain the super-resolution (SR) [45], the bicubic method [46], and plug-and-play alternating direction multiplier method (PAP-ADMM) [47].

This chapter is organized as follows. In Section 2, we review the basic idea of color mapping and its variants. In Section 3, we include extensive experiments to illustrate the performance of various pansharpening algorithms. Finally, we conclude our chapter with some future research directions.

2. Proposed hybrid color mapping algorithm and its variants

2.1. Basic idea of color mapping for generating HR HS images

As shown in **Figure 1**, the idea of color mapping is to map a color pixel $c_{(i,j)}$ at location (i, j) with R, G, B bands to a hyperspectral pixel $X_{(i,j)}$ at the same location. This mapping is represented by a transformation matrix T, that is

$$X_{(i,j)} = Tc_{(i,j)} \tag{1}$$

where $X_{(i,j)} \in R^N$ is a single hyperspectral pixel with N spectral bands, $T \in R^{N\times M}$, $c_{(i,j)} \in R^M$ is a color pixel with M spectral bands, and $N>>M$. Here, M can be just one band such as the pan band. Hence, color mapping is quite general as it encompasses pan, color, and MS images. Our goal is to generate a HR HS image given a HR color image and an LR HS image. To determine T in Eq. (1), we simulate an LR color image by down-sampling the HR color image. The LR color image and the LR HS image are then used to determine T, which is then used to generate the HR HS image pixel by pixel.

Let us denote H as the set of all hyperspectral pixels $X_{(i,j)}$ for all (i,j) in the image and C as the set of all color pixels $c_{(i,j)}$ for all (i,j) in the image, $i = 1, \ldots, N_R$, $j = 1, \ldots, N_C$ with N_R the number of rows and N_C the number of columns in the image. Since $X_{(i,j)}$ and $c_{(i,j)}$ are vectors, H and C can be expressed as

$$H = \left[X_{(1,1)}\ X_{(1,2)}\ \ldots\ X_{(N_R,N_C)}\right],\ C = \left[C_{(1,1)}\quad C_{(1,2)}\quad \ldots\quad C_{(N_R,N_C)}\right]. \tag{2}$$

We call the mapping (1) the global version and all pixels in C and H are used in estimating T.

To estimate T, we use the least-square approach, which minimizes the error

$$E = \|H - TC\|_F^2. \tag{3}$$

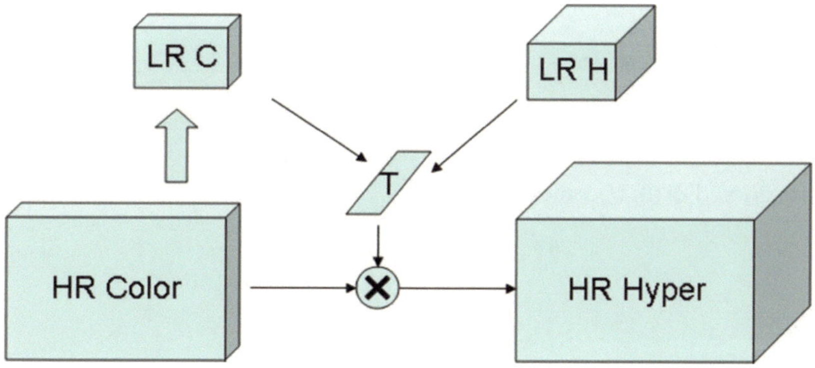

Figure 1. System flow of color mapping. LR denotes low resolution; HR denotes high resolution; LR C denotes the set of low-resolution color pixels; LR H denotes the set of low-resolution hyperspectral pixel; HR Hyper denotes high-resolution hyperspectral.

Solving T in Eq. (3) yields [48]

$$T^{*} = HC^{T}\left(CC^{T}\right)^{-1}. \tag{4}$$

To avoid instability, we can add a regularization term in Eq. (3). That is,

$$T^{*} = \arg\min_{T} \|H - TC\|_{F} + \lambda\|T\|_{F}. \tag{5}$$

And the optimal T becomes [48]

$$T^{*} = HC^{T}\left(CC^{T} + \lambda I\right)^{-1} \tag{6}$$

where λ is a regularization parameter and I is an identity matrix with the same dimension as CC^{T}.

2.2. Hybrid color mapping

For many hyperspectral images, the band wavelengths range from 0.4 to 2.5μm. For color images, the R, G, and B wavelengths are 0.65, 0.51, and 0.475μm, respectively. Because the three color bands may have little correlation with higher-number bands in the hyperspectral image, we found that it is beneficial to extract several higher-number bands from LR HS image and stack them with the LR color bands. This idea is illustrated in **Figure 2**. Details can be found in [48]. Moreover, we also noticed that by adding a white band, that is, all pixel values are 1, we can deal with atmospheric and other bias effects.

Using the same treatment earlier, T can be obtained by minimizing the mean square error [48]

$$T^{*} = \arg\min_{T} \|H - TC_{h}\|_{F} \tag{7}$$

where H is the set of hyperspectral pixels and C_h is the set of hybrid color pixels. All the pixels in H and C_h are used.

The optimal T can be determined as

$$T^{*} = HC_{h}^{\ T}\left(C_{h}C_{h}^{\ T}\right)^{-1} \tag{8}$$

With regularization, Eq. (8) becomes [48]

$$T^{*} = HC_{h}^{\ T}\left(C_{h}C_{h}^{\ T} + \lambda I\right)^{-1}. \tag{9}$$

2.3. Local HCM

We further enhance our method by applying color mapping patch by patch. As shown in **Figure 3**, a patch of size $p \times p$ is a sub-image in the original image. The patches can be overlapped. This idea allows spatial correlation to be exploited. Our experiments showed that the mapping will become more accurate using this local patch idea. Another advantage of using patches is to split the task into many small jobs so that parallel processing is possible.

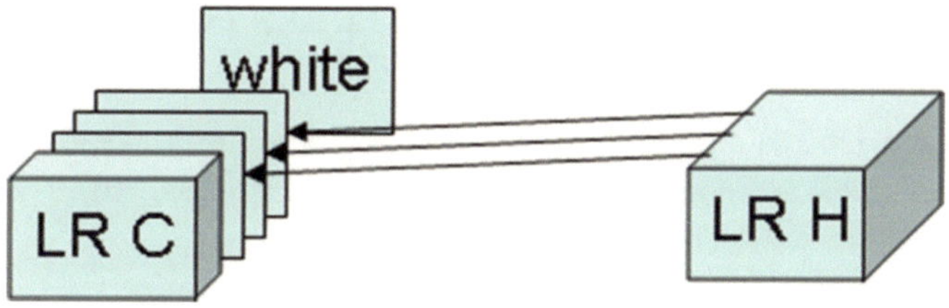

Figure 2. System flow of hybrid color mapping.

2.4. Incorporation of PSF into HCM

Based on observations in [28] and our own investigations [48], some pansharpening algorithms with the incorporation of PSF can yield better performance than those without PSF. This motivates to incorporate PSF into our HCM. Our idea can be illustrated by using **Figure 4**. The first component is the incorporation of a single-image super-resolution algorithm to enhance the LR hyperspectral image cube. Single-image super-resolution algorithms are well known [45, 46]. The idea is to improve the resolution of an LR image by using internal image statistics. Our proposed method enhances the LR HS bands and then fuses the results using the HCM algorithm. The second component utilizes the HCM algorithm that fuses a high-resolution color image with an enhanced hyperspectral image coming out of the first component. Recently, HCM has been applied to several applications, including enhancing

Figure 3. Local color mapping to further enhance the SR performance. The patches apply to LR color and LR hyperspectral images.

Worldview-3 images [49], fusion of Landsat and MODIS images [50], pansharpening of Mastcam images [51], and fusing of THEMIS and TES [52].

This idea was summarized in a paper [53] that was presented in ICASSP 2017. The results are comparable to other state-of-the-art algorithms in the literature [39, 40].

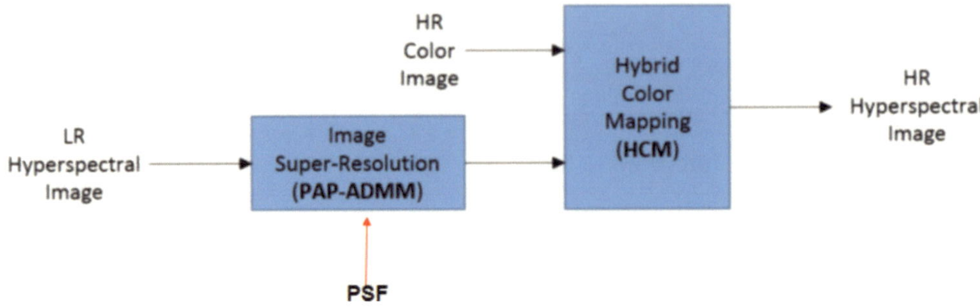

Figure 4. An outline of the proposed method. We use hybrid color mapping (HCM) to fuse low-resolution (LR) and high-resolution (HR) images. For LR images, we use a single-image super-resolution algorithm where PSF is incorporated to first enhance the resolution before feeding to the HCM.

Figure 5. Standard Bayer pattern.

2.5. Sparsity-based variants using L_1 and L_0 norms

In [54], we propose two variants of HCM. From Eq. (5), we can treat the HCM method as an L_2 regularization problem. In [54], we investigate some variants of the regularization. In particular, we would like to apply L_1 and L_0 norms for the regularization term in Eq. (5). We favor the following two approaches: orthogonal matching pursuit (OMP) [55] and l1-minimization via augmented Lagrangian multiplier (ALM-l1) [56], in which OMP is described as an l0-based minimization problem:

$$T^* = \arg\min_T \|H - TC\|_F \quad \text{s.t.} \|T\|_0 \le K \tag{10}$$

where K is the sparsity level of the matrix α ($K << M \times N$), and ALM-l1 solves for the l1-minimization convex relaxation problem:

$$T^* = \arg\min_T \|H - TC\|_F + \lambda \|T\|_1 \tag{11}$$

where the positive-weighting parameter λ provides the trade-off between the two terms.

2.6. Application of pansharpening algorithms to debayering

Debayering or demosaicing refers to the reconstruction of missing pixels in the Bayer pattern [57, 58] as shown in **Figure 5**. A variant of the Bayer pattern, known as CFA2.0 (**Figure 6**), was introduced in [59]. Recently, a new approach [60] to debayering was introduced based on pansharpening ideas. A

Figure 6. RGBW (aka CFA2.0) pattern.

thorough comparative study was performed using benchmark datasets. It was found that the pansharpening-based debayering approach holds good promise.

3. Experiments

3.1. Data

Two hyperspectral image datasets were used in our experiments. One was from the Air Force (AF) and the other one was one of the NASA AVIRIS' images. The AF image (**Figure 7**) has 124 bands (461–901 nm). The AVIRIS image (**Figure 8**) has 213 bands (380–2500 nm). Both of them are natural scenes. The AF image size is 267×342×124 and the AVIRIS image size is 300×300×213.

The down-sampled image was used as the low-resolution hyperspectral image that needs to be improved. We picked R, G, and B bands from the original high-resolution hyperspectral image for color mapping. The bicubic method in the following plots was implemented by up-sampling the low-resolution image using bicubic interpolation. The results of bicubic method were used as baseline for comparison study.

3.2. Performance metrics

Similar to [28], five performance metrics are included here.

Time: It is the computational time in seconds. This metric is machine dependent and varies with runs. However, it gives a relative measure of the computational complexity of different algorithms.

Figure 7. Sample band of AF data.

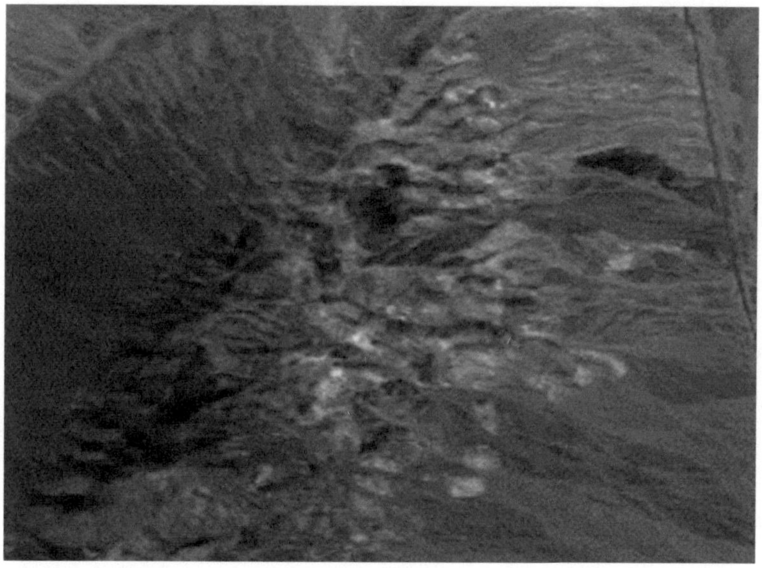

Figure 8. Sample band of AVIRIS data.

Root mean-squared error (RMSE) [28]: Given two matrices X and $\widehat{X}$, the RMSE is calculated by using

$$RMSE\left(X,\widehat{X}\right) = \frac{\left\| X\text{-}\widehat{X} \right\|_F}{\sqrt{\text{total number of pixels}}}. \tag{12}$$

The ideal value of RMSE is 0 if there is perfect reconstruction. To show the performance for each band, we also used $RMSE(\lambda)$, which is the RMSE value between $X(\lambda)$ and $\widehat{X}(\lambda)$ for each band λ.

Cross-correlation (CC) [28]: It is defined as

$$CC\left(X,\widehat{X}\right) = \frac{1}{m_\lambda}\sum_{i=1}^{m_\lambda} CCS\left(X^i,,\widehat{X}^i\right) \tag{13}$$

where m_λ is the number of bands in the hyperspectral image and CCS is the cross-correlation for a single-band image, given by

$$CCS(A,B) = \frac{\sum_{j=1}^{n}\left(A_j-\mu_A\right)\left(B_j-\mu_B\right)}{\sqrt{\sum_{j=1}^{n}\left(A_j-\mu_A\right)^2\sum_{j=1}^{n}\left(B_j-\mu_B\right)^2}}. \tag{14}$$

The ideal value of CC is 1. We also used $CC(\lambda) = CCS\left(X^i,,\widehat{X}^i\right)$, which is the CC value between $X(\lambda)$ and $\widehat{X}(\lambda)$ for each band, to evaluate the performance of different algorithms.

Spectral Angle Mapper (SAM) [28]: It is defined as

$$SAM\left(X,\widehat{X}\right)=\frac{1}{n}\sum_{j=1}^{n}SAM\left(x_j,\widehat{x}_j\right) \tag{15}$$

where, for two vectors $a,b\in R^{m_\lambda}$,

$$SAM(a,b)=\arccos\left(\frac{a,b}{\|a\|\|b\|}\right). \tag{16}$$

$\langle a,b\rangle$ is the inner product between two vectors and $\|\bullet\|$ denotes the two norms of a vector. The ideal value of SAM is 0.

Erreur relative globale adimensionnelle de synthèse (ERGAS) [28]: It is defined as

$$ERGAS\left(X,\widehat{X}\right)=100d\sqrt{\frac{1}{m_\lambda}\sum_{k=1}^{m_\lambda}\left(\frac{RMSE_k}{\mu_k}\right)^2} \tag{17}$$

where d is the ratio between the linear resolutions of the PAN and HS images, μ_k is the sample mean of the kth band of X. The ideal value of ERGAS is 0.

3.3. Advantages of the local mapping approach as compared to the global mapping method

We first compared global color mapping (Section 2.1) with local color mapping (Section 2.2) and bicubic interpolation method. **Figures 9** and **10** show the RMSE between the ground-truth hyperspectral images and the super-resolution images produced by different methods. We can see that local color mapping is always better than global color mapping. In the AF image, for lower-number bands where R, G, and B bands reside, both global and local color mapping produce better performance than bicubic method. See **Figure 9**. However, for higher-number bands, bicubic method is better. The reason is that the spectral correlation between higher-number bands and R, G, B bands is weak. For the AVIRIS image, the local color mapping results shown in **Figure 10** are better than both global color mapping and bicubic results across almost all spectral bands.

3.4. Advantages of hybrid color mapping

Figures 11 and **12** show the performance of hybrid color mapping as well as RGBW (R, G, B, and white bands) color mapping described in Section 2. *W* refers to the white band, that is, an image with all 1s. We can see that adding a white band improves the performance across all bands. Moreover, the hybrid method, which combines more bands from the bicubic interpolated higher-number bands, performed the best in all of the bands. Here, all methods used local patches. The patch size is 4 × 4 and there is no overlapping between the patches.

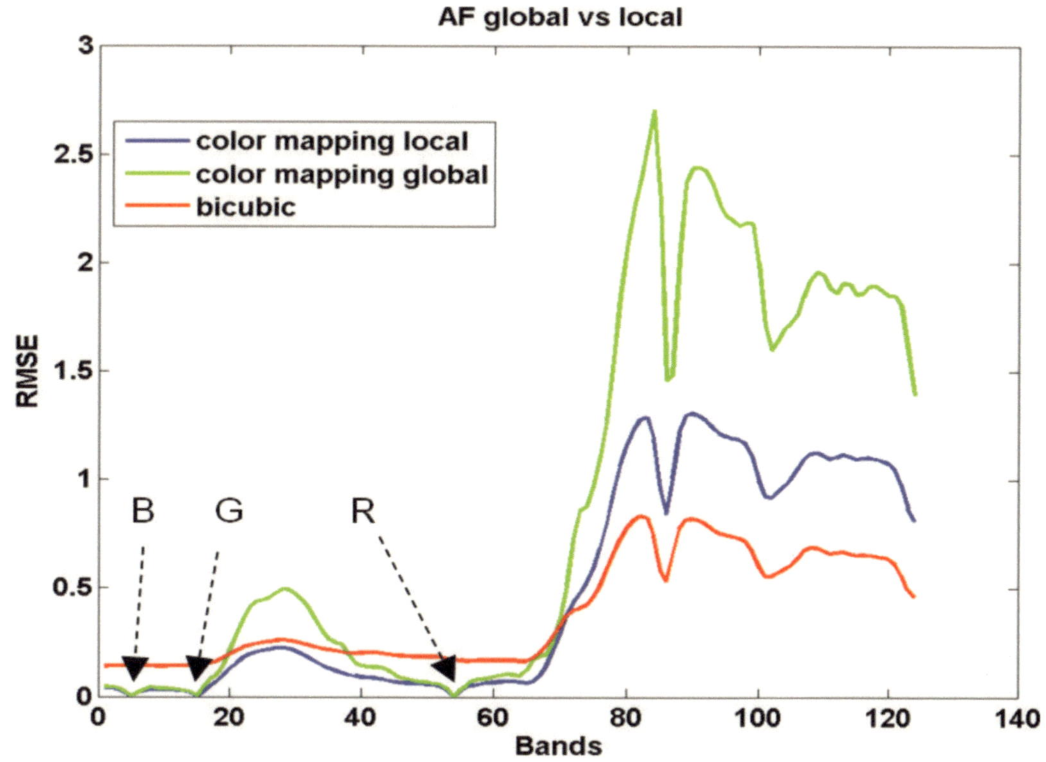

Figure 9. RMSE comparison for AF dataset.

3.5. Comparison with state-of-the-art algorithms

Now, we include a comparison with a state-of-the-art pan-sharpening algorithm [61] and another single-image SR algorithm [45]. Both algorithms are recent and compared favorably than their counterparts in their categories. **Figures 13** and **14** show the performance of variational wavelet pan-sharpening (VWP) [61], bicubic interpolation, single-image SR [45], and our local hybrid color mapping. We used red color band from ground-truth image as the pan image. For NASA data, we observe that in some bands, VWP is better than bicubic and single-image SR methods. However, for bands far away from the reference band, the error is large. In addition, other methods are always worse than our hybrid color mapping method. The reason that VWP [61] did not perform well in this study is perhaps due to the lack of a pan image which has a spectrum that extends to the high-wavelength regions. If the pan image extends to higher-wavelength regions, it is believed that VWP will perform well. We are grateful to the authors of [45] for sharing their source codes with us. The reason that the single-image SR [45] method did not perform well is because it was designed for color images, not for hyperspectral images.

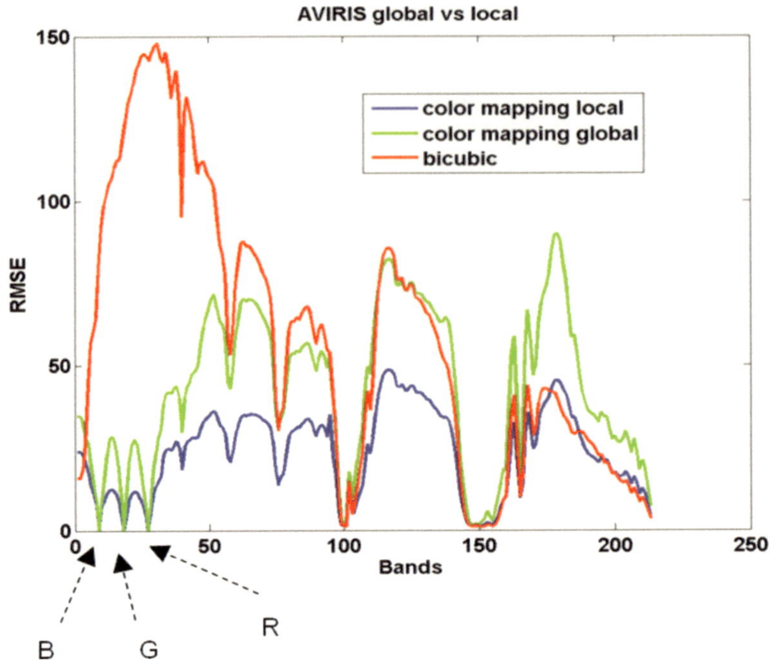

Figure 10. RMSE comparison for AVIRIS dataset.

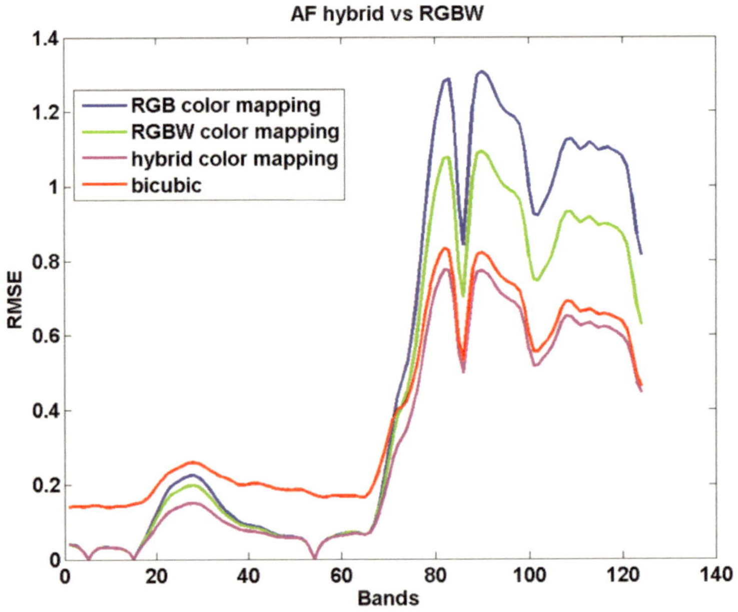

Figure 11. RGB color mapping versus hybrid color mapping for AF dataset. *W* stands for white band. All methods are local. RGBW means a white band is added to the R, G, B bands.

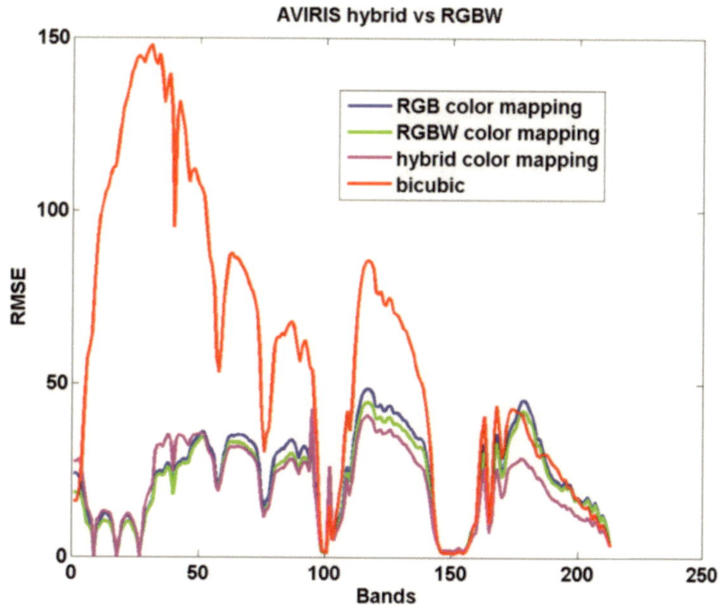

Figure 12. RGB color mapping versus hybrid color mapping for AVIRIS dataset. *W* stands for white band. All methods are local. RGBW means a white band is added to the R, G, B bands.

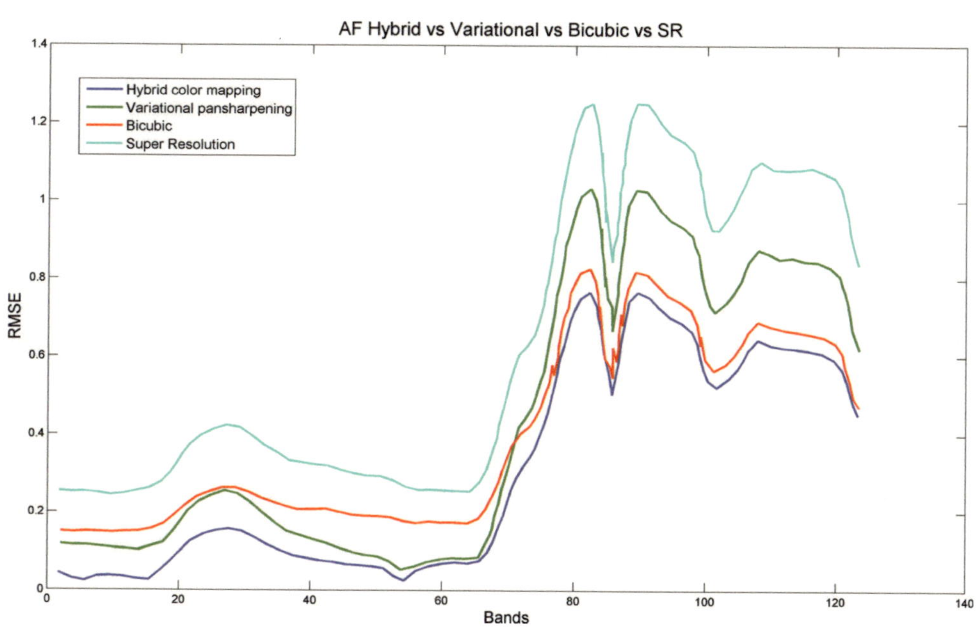

Figure 13. Hybrid color mapping versus variational wavelet pan-sharpening (VWP), bicubic, and single-image SR [45] for AF dataset. Scale factor is 3.

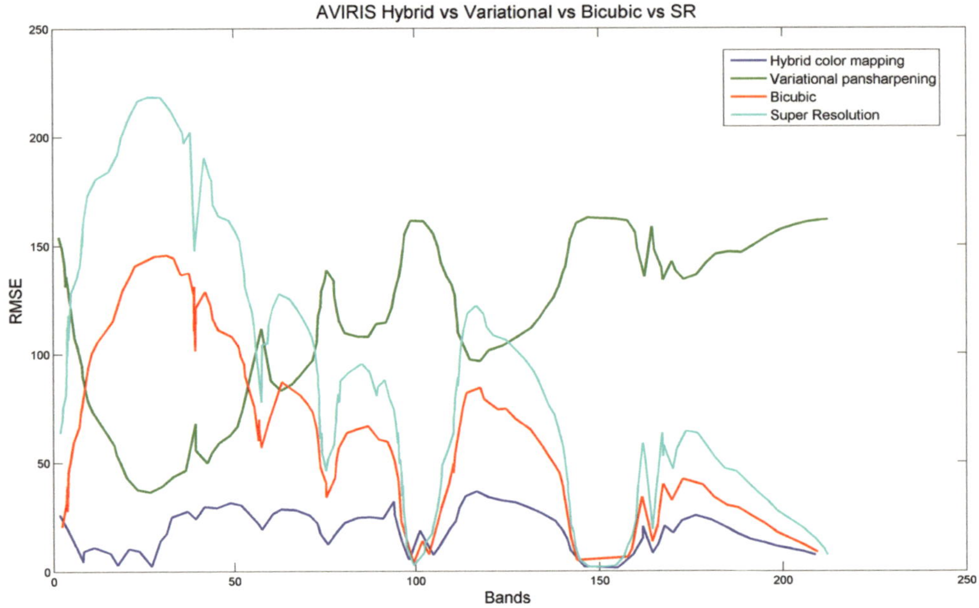

Figure 14. Hybrid color mapping versus variational wavelet pan-sharpening (VWP), bicubic, and single-image SR [45] for AF dataset. Scale factor is 3.

3.6. Pixel clustering enhancement using the SR images

The goal of our research is not only to improve the visual performance of the hyperspectral images by enhancing the spatial resolution but also to enhance the pixel clustering performance as well. **Figures 15–18** show clustering results using end members extracted from ground-truth AVIRIS hyperspectral image. We used K-means end member extraction technique to determine the clusters. It should be emphasized that we are not performing land-cover classification, which usually requires land-cover spectral signatures as well as atmospheric compensation. The physical meaning of each cluster is not our concern. Comparing **Figures 15** and **16**, one can see that hybrid color mapping is significantly better than the bicubic method in terms of the fine details of the images. Moreover, **Figure 16** is a lot closer to the ground-truth image in **Figure 17**. The images also show that the hybrid color mapping produces much better clustering than the bicubic method as shown in **Figure 18**.

3.7. Additional studies and observations

It should be noted that in all of the above studies, we have the following observations:

- In the AVIRIS image, our local hybrid color mapping algorithm performed consistently well in all bands. That is, our method is better than bicubic and other methods in all bands.
- In the AF image, our local hybrid color mapping algorithm performed also better than others across all bands. However, the performance in lower-number bands is better than that in higher-number bands. This is because the correlation between lower-number bands and higher-number bands is not that good.

- We are curious about what will happen if bands other than R, G, B are chosen as reference image in generating the transformation T. **Figures 19** and **20** show the comparison. **Figure 19** compares the RMSEs using only R, G, B and the case of using bands 20, 50, and 100, for the AF image. **Figure 20** compares the RMSEs using only RGB and the case of using bands 20, 90, and 170, for the AVIRIS image. We can observe that if we pick bands from low-, middle-, and high-number bands assuming that they are available, then the overall performance can be much better than that of using R, G, B bands only. The observation shows that we should utilize not only R, G, B bands information but also other bands information such as infrared image and thermal image if they are available.

- We investigated the impact of adding regularization to the solution of finding T. In our simulations, we used $\lambda = 10^{-5}v$ and v is the maximum singular value of CC^T in Eq. (5) or $C_hC_h^T$ in Eq. (8). From **Figures 21** and **22**, it can be seen that regularization can also further enhance the performance of our algorithm.

- We also investigated the impact of different scaling factors in the down-sampling process. As can be seen in the supplemental materials section, results indicate that bicubic method performs well in low-scaling factors of 1.5 where spatial correlation is high. However, in our applications, we are interested in airborne and satellite images where spatial correlation is weak.

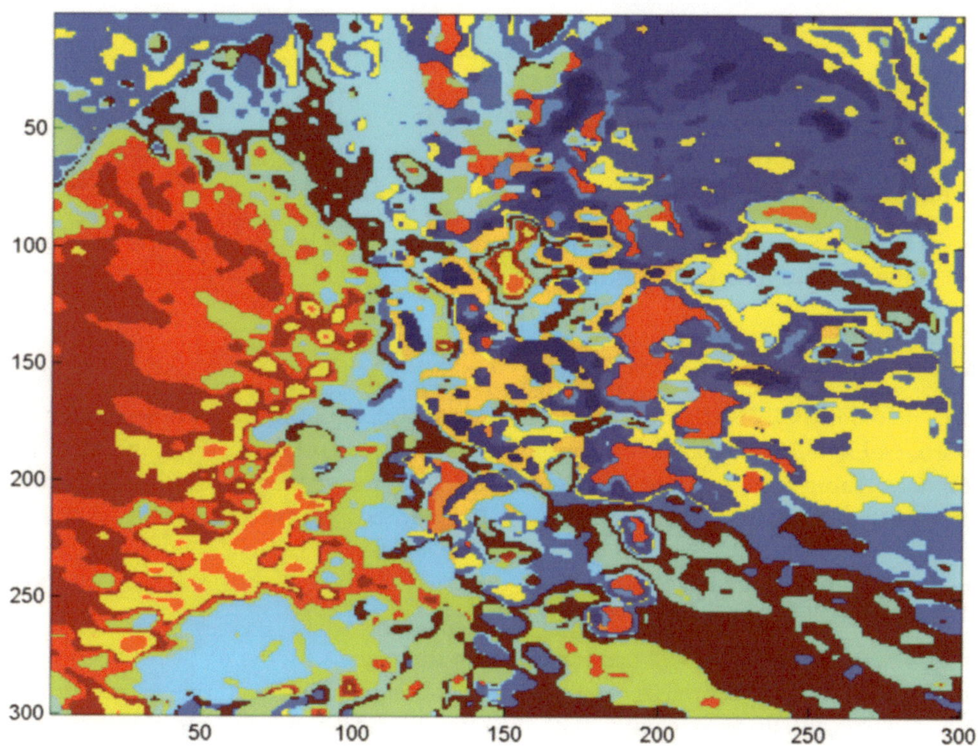

Figure 15. Bicubic pixel classification.

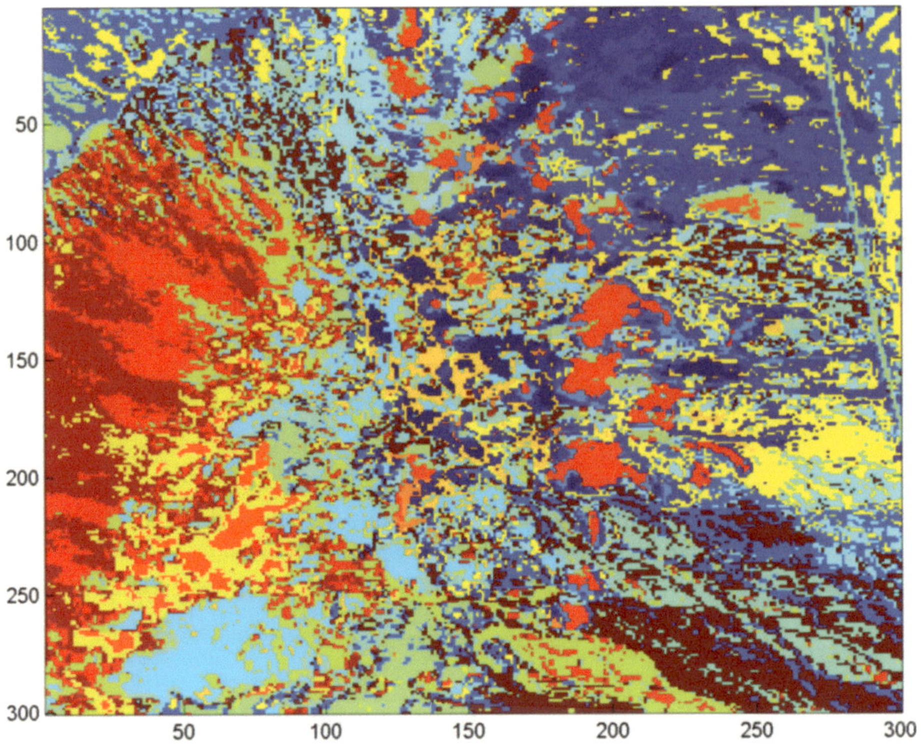

Figure 16. Hybrid color mapping pixel classification.

3.8. Comparison of sparsity-based approach with other methods

Here, we focus on a comparison with Groups 1–3 algorithms in the literature. We also compare the performance of different variants of HCM.

3.8.1. Comparison with Group 1 methods

Group 1 methods (BN, BS, CNMF) require PSF. In general, Group 1 methods performed better than Groups 1 and 2. In Section 2.4, we also incorporated PSF into our HCM algorithm. Based on results in **Table 1**, we can see that the HCM (L2-norm) yielded better results than those Group 1 methods for the AF data. For the AVIRIS data, BS performed the best.

3.8.2. Comparison with Group 2 methods

Only the GSA method in Group 2 performed better than others for the AVIRIS. Other methods in Group 2 did not perform well as compared to HCM and Group 2 methods. However, Group 2 methods are generally more computationally efficient, which may be advantageous in some real-time applications.

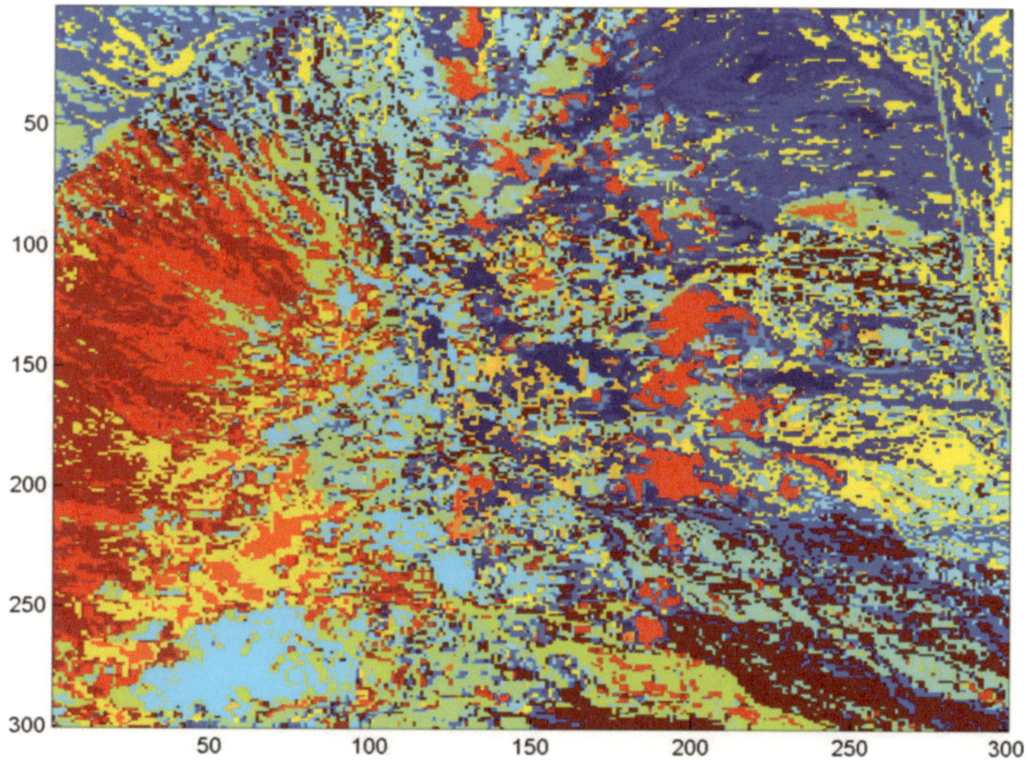

Figure 17. Ground-truth pixel classification.

3.8.3. Comparison with Group 3 methods

Group 3 methods do not require PSF or high-resolution color images. Consequently, the performance is generally poor as compared to others. This is understandable.

3.8.4. Comparison among the HCM variants

From **Table 1**, we can see that the L_2 version of HCM performed better than other variants. This can also be seen from **Figures 23–26**. However, one key advantage of the L_1 and L_0 variants is that the models are simpler, as the coefficients of L_0 are clustered and the coefficients of L_1 are much fewer than that of L_2. The above can be confirmed by inspecting **Figures 27** and **28**. We would like to mention that another advantage of the sparsity formulation is that it can handle noisy measurements in the hyperspectral images [54].

4. Conclusions

In this chapter, we review the various pansharpening algorithms for hyperspectral images. We then focus on one recent algorithm known as hybrid color mapping. Several variants are

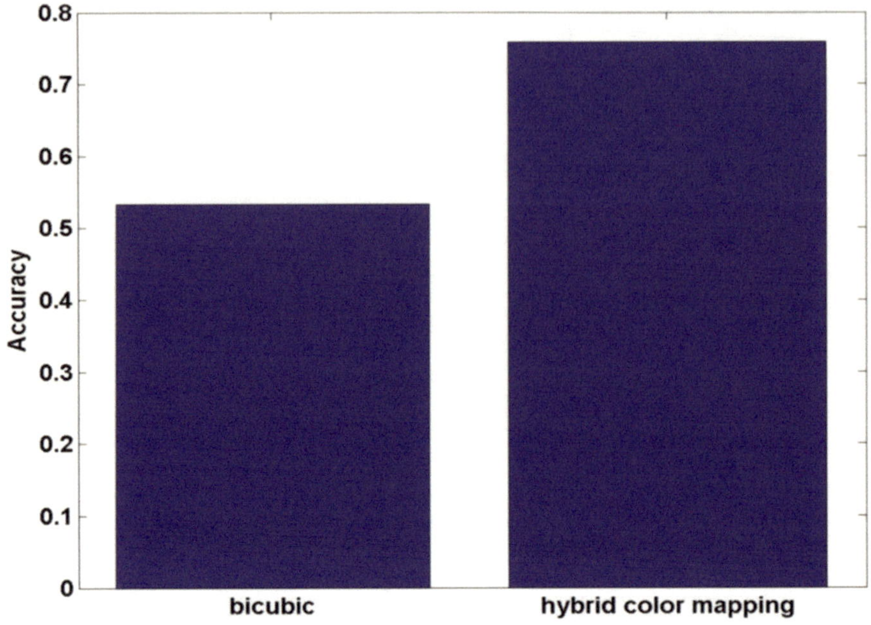

Figure 18. Pixel classification accuracy.

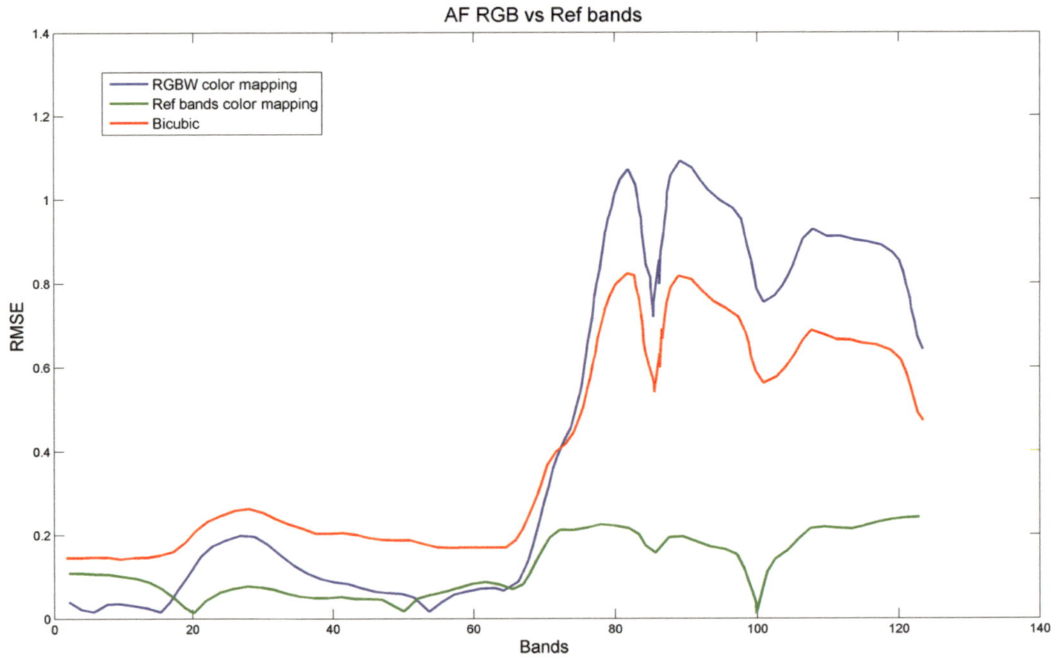

Figure 19. Local color mapping with RGBW bands versus local color mapping with low-, middle-, and high-number bands for AF dataset.

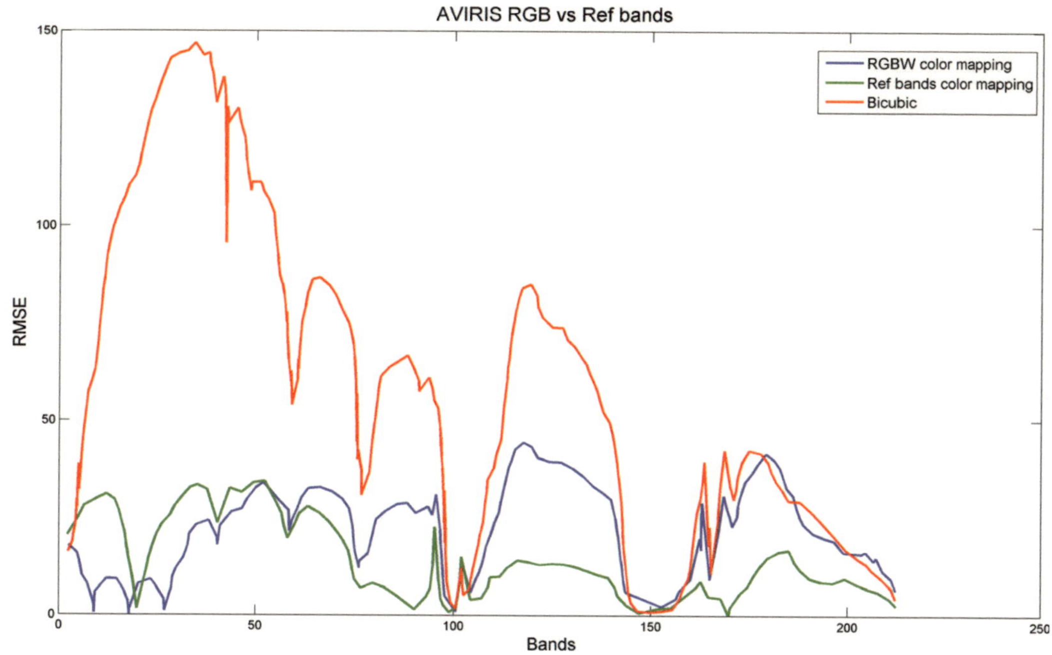

Figure 20. Local color mapping with RGBW bands versus local color mapping with low-, middle-, and high-number bands for AVIRIS dataset.

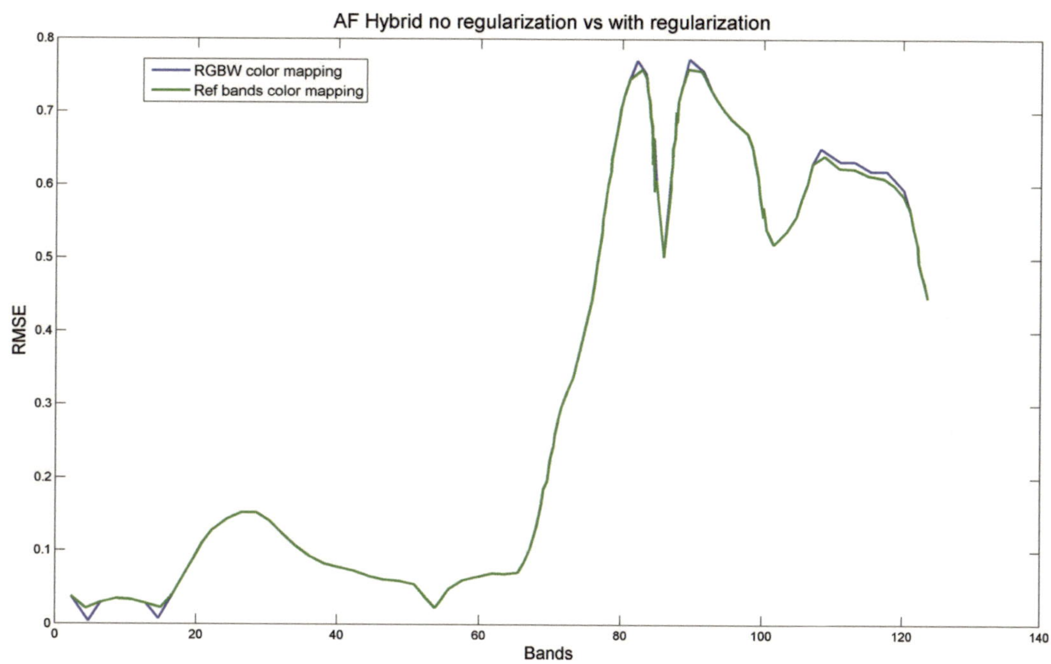

Figure 21. Regularization improves the stability as well as the performance. Here, local hybrid color mapping has been used for AF dataset.

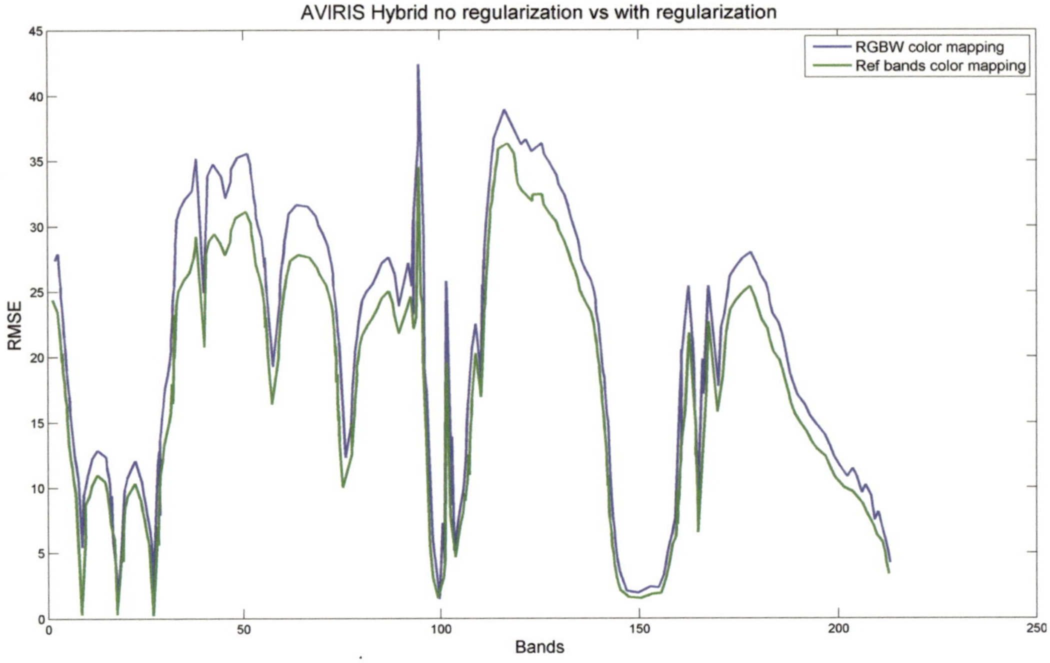

Figure 22. Regularization improves the stability as well as the performance. Here, local hybrid color mapping has been used for AVIRIS dataset.

		AF					AVIRIS				
Group	Methods	Time	RMSE	CC	SAM	ERGAS	Time	RMSE	CC	SAM	ERGAS
1	CNMF [34]	12.52	0.5992	0.9922	1.4351	1.7229	23.75	32.2868	0.9456	0.9590	2.1225
	Bayes Naive [32]	0.58	0.4357	0.9881	1.2141	1.6588	0.86	67.2879	0.9474	0.8136	2.1078
	Bayes Sparse [33]	208.82	0.4133	0.9900	1.2395	1.5529	235.50	51.7010	0.9619	**0.7635**	1.8657
2	SFIM [30]	0.99^+	0.7132	0.98489	1.4936	2.2087	1.56y	62.02333	0.94911	0.918	2.0404
	MTF-GLP [26]	1.38^+	0.8177	0.98314	1.6095	2.4541	2.25y	55.604	0.95451	0.91031	1.9703
	MTF-GLP-HTM [27]	1.40^+	0.8050	0.98356	1.546	2.4214	2.23y	55.641	0.95451	0.90544	1.9718
	GS [24]	1.05^+	2.1783	0.85792	2.4421	7.0807	1.83y	54.8851	0.95597	0.93488	1.9524
	GSA [25]	1.21^+	0.7435	0.98764	1.5156	2.1764	1.98y	32.2331	**0.97016**	0.85211	**1.6525**
	PCA [23]	2.37^+	2.3816	0.83824	2.6355	7.7176	2.98y	48.9125	0.96087	0.9173	1.8605
	GFPCA [31]	1.17^+	0.6482	0.98615	1.5382	2.0607	2.17y	62.5283	0.93858	1.1736	2.2566
	Hysure [28, 29]	117.06^+	0.8717	0.98059	1.7882	2.6294	62.47y	38.3131	0.9598	1.0181	1.8586
3	PAP-ADMM [42]	2144.00	0.4408	0.98849	1.1657	1.6476	3368.00	66.2481	0.9531	0.7848	1.9783

		AF					AVIRIS				
	Super-Resolution [39]	279.18	0.5232	0.9839	1.3215	1.9584	1329.59	86.7154	0.9263	0.9970	2.4110
	Bicubic [40]	**0.04**	0.5852	0.9807	1.3554	2.1560	**0.10**	92.2143	0.9118	1.0369	2.5728
Ours	L_2 Norm	3.70[+]	**0.3889**	**0.9965**	**1.1928**	**1.0986**	4.47	**32.2094**	0.9553	0.9453	1.9101
	L_0 Norm	33.36[+]	0.4288	0.9958	1.3677	1.2095	81.11	32.1875	0.9552	0.9362	1.9109
	L_1 Norm	1293.94[+]	0.3917	0.9965	1.2259	1.1064	6.22	32.1875	0.9552	0.9362	1.9109

[+]These methods involve PAP-ADMM but we did not include PAP-ADMM's runtime in order to illustrate the differences. Bold numbers indicate results from the best algorithms.

Table 1. Comparison of our methods with various pansharpening methods on AF and AVIRIS.

described. The performance of HCM is thoroughly compared with other methods in the literature.

One future direction is to investigate the performance of different pansharpening algorithms in the presence of noise. Another direction is to apply the pansharpened images to different applications such as target detection [11–13], border monitoring [24, 25], and anomaly detection [14–22, 62].

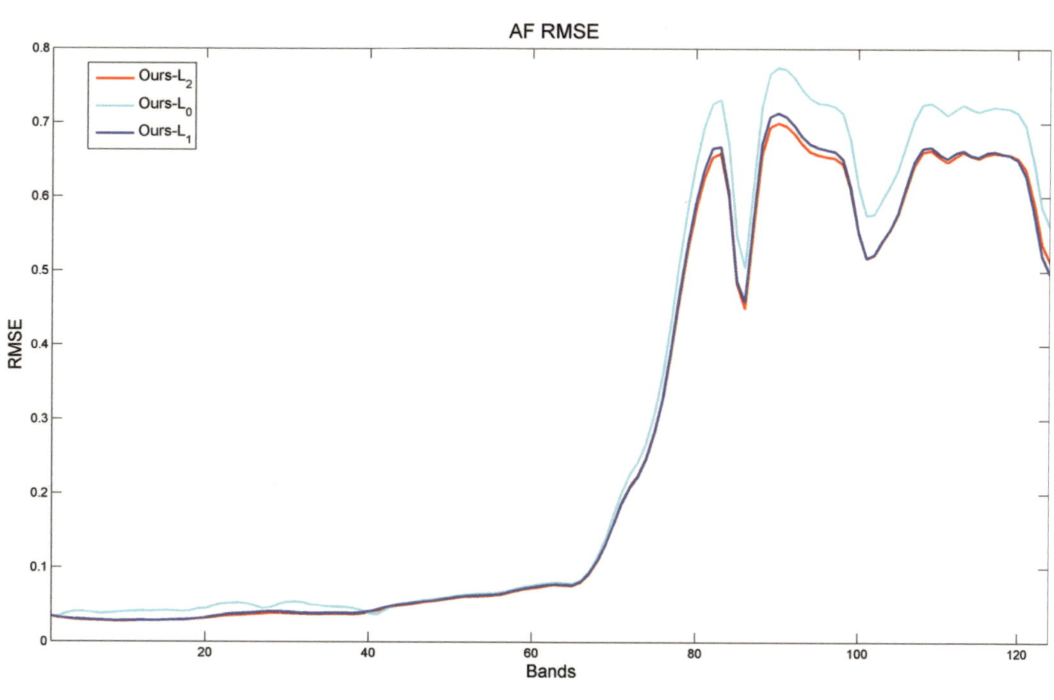

Figure 23. RMSE comparison of the three variants of HCM (L2, L1, and L0) for the AF dataset.

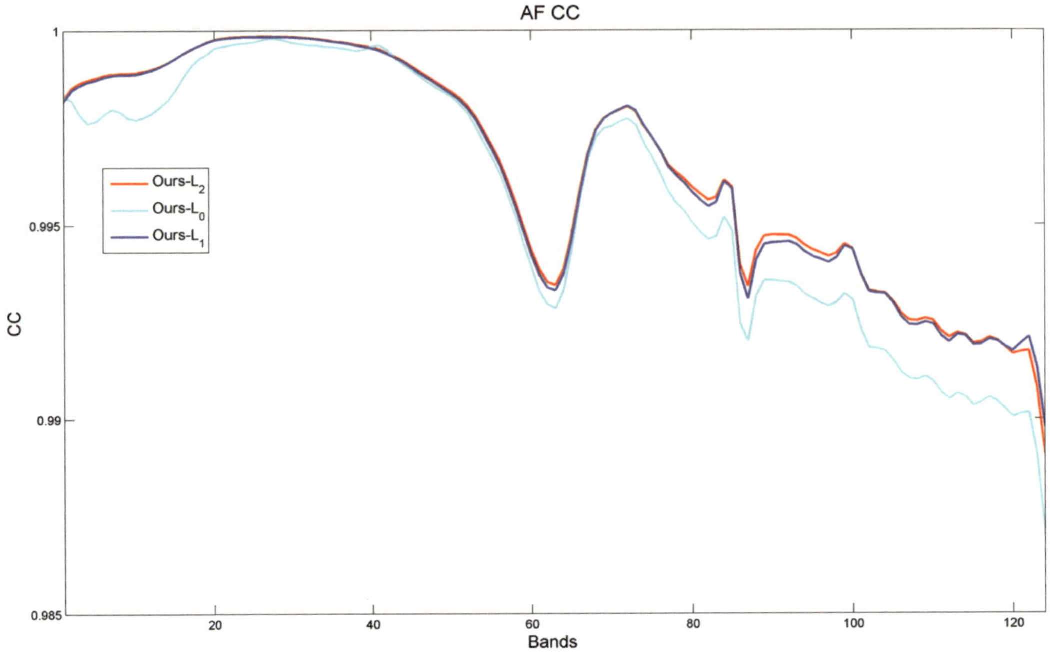

Figure 24. CC comparison of the three variants of HCM (L2, L1, and L0) for the AF dataset.

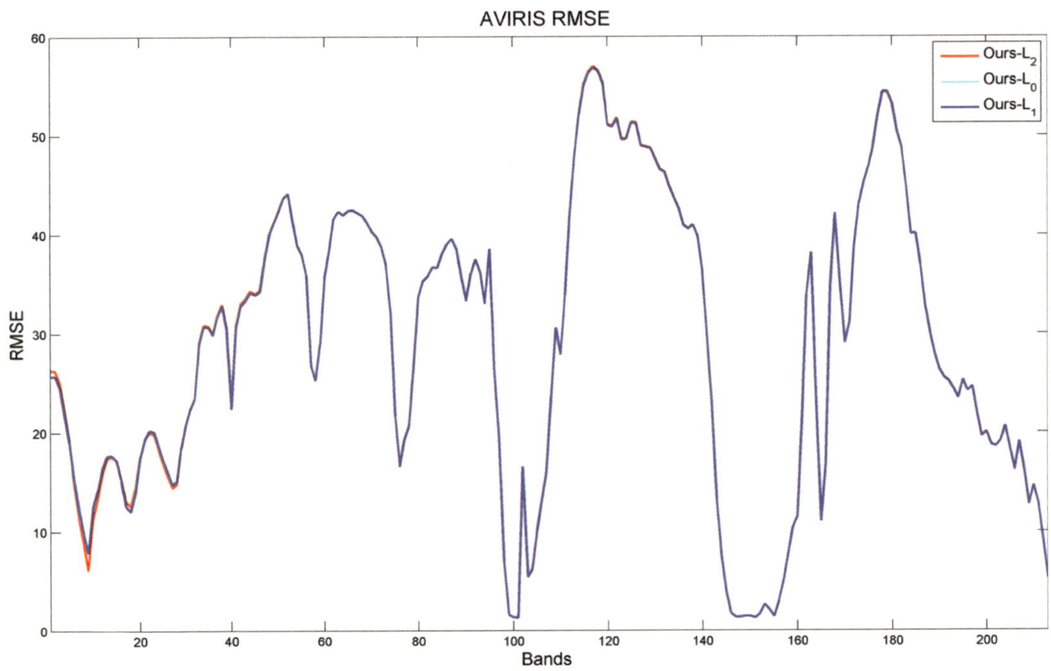

Figure 25. RMSE comparison of the three variants of HCM (L2, L1, and L0) for the AVIRIS dataset.

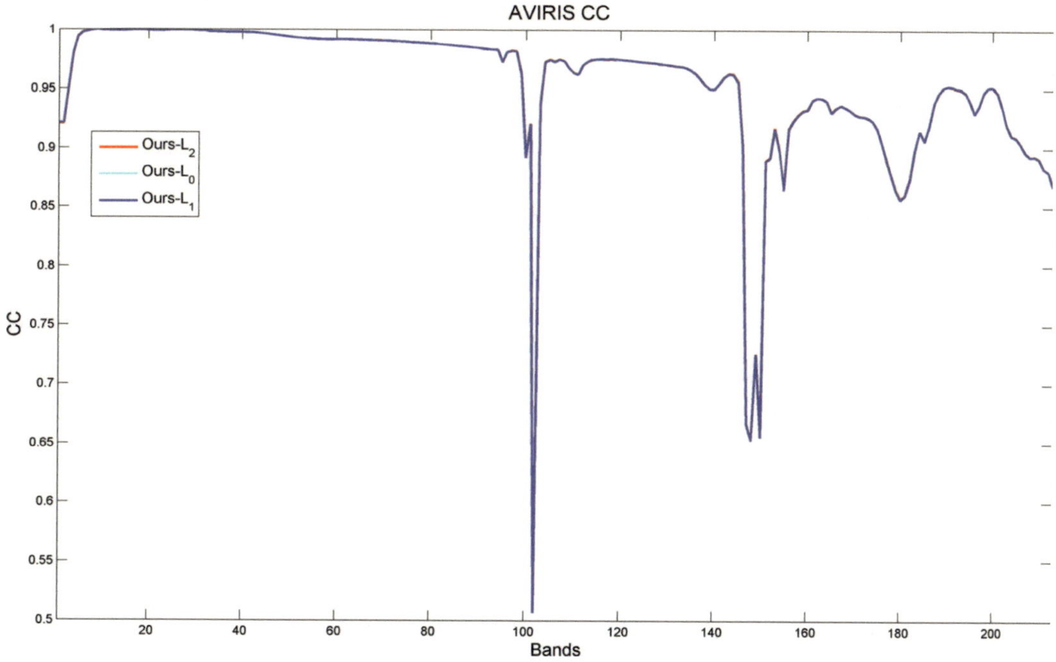

Figure 26. CC comparison of the three variants of HCM (L2, L1, and L0) for the AVIRIS dataset.

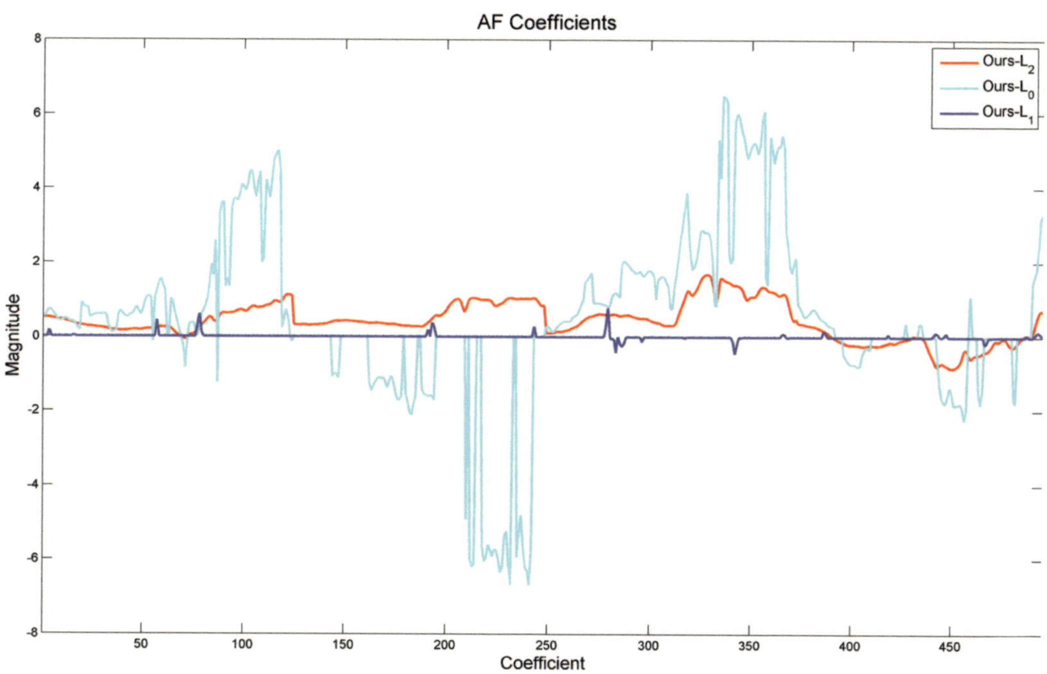

Figure 27. Coefficients in the model of the HCM variants for the AF dataset.

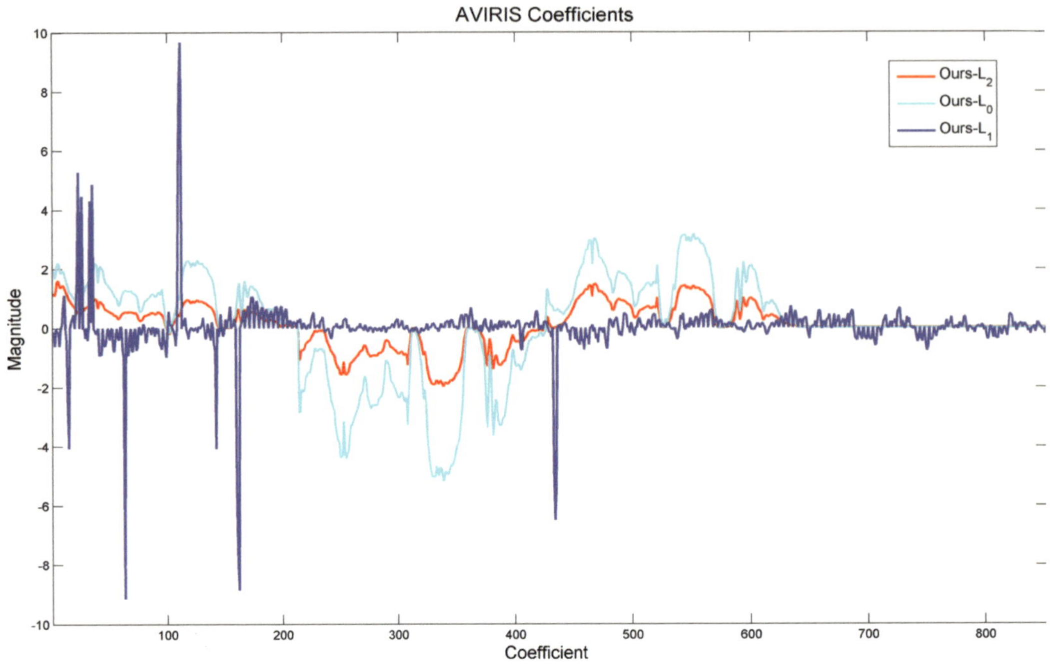

Figure 28. Coefficients in the model of the HCM variants for the AVIRIS dataset.

Author details

Chiman Kwan*, Jin Zhou and Bence Budavari

*Address all correspondence to: chiman.kwan@signalpro.net

Signal Processing, Inc., Rockville, MD, USA

References

[1] Zhou J, Chen H, Ayhan B, Kwan C. A high performance algorithm to improve the spatial resolution of HyspIRI images. In: NASA HyspIRI Science and Applications Workshop. Washington DC: NASA Jet Propulsion Laboratory; 2012 Oct

[2] Ayhan B, Zhou J, Kwan C. High performance and accurate change detection system for HyspIRI missions. In: NASA HyspIRI Science Symposium, Greenbelt. Maryland: NASA Jet Propulsion Laboratory; 2012 May 16

[3] Kwan C, Yin J, Zhou J, Chen H, Ayhan B. Fast parallel processing tools for future HyspIRI data Processing. In: NASA HyspIRI Science Symposium. Greenbelt. Maryland: NASA Jet Propulsion Laboratory; 2013 April

[4] Ayhan B, Kwan C. Fast target detection framework for onboard processing of multispectral and hyperspectral images. In: HyspIRI Science Symposium. Greenbelt. Maryland: NASA Jet Propulsion Laboratory; 2015 Jun 3

[5] CRISM [Internet]. Available from: http://crism.jhuapl.edu/index.php [Accessed: Sep 6, 2017]

[6] Zhou J, Kwan C, Ayhan B. A high performance missing pixel reconstruction algorithm for hyperspectral images. In: 2nd Int. Conf. on Applied and Theoretical Information Systems Research. Taipei: Academy of Taiwan Information Systems Research; 2012 Dec 27

[7] Nguyen D, Tran T, Kwan C, Ayhan B. Endmember extraction in hyperspectral images using l-1 minimization and linear complementary programming. In Proceedings of SPIE. Orlando: SPIE; 2010 Apr 23 (Vol. 7695, pp. 76951-76951)

[8] Kwan C, Ayhan B, Chen G, Wang J, Ji B, Chang CI. A novel approach for spectral unmixing, classification, and concentration estimation of chemical and biological agents. IEEE Transactions on Geoscience and Remote Sensing. 2006 Feb;**44**(2):409-419

[9] Kwan C, Schmera G, Smulko JM, Kish LB, Heszler P, Granqvist CG. Advanced agent identification with fluctuation-enhanced sensing. IEEE Sensors Journal. 2008 Jun;**8**(6):706-713

[10] Kish LB, Chang HC, King MD, Kwan C, Jensen JO, Schmera G, Smulko J, Gingl Z, Granqvist CG. Fluctuation-enhanced sensing for biological agent detection and identification. IEEE Transactions on Nanotechnology. 2011 Nov;**10**(6):1238-1242

[11] Ayhan B, Kwan C. On the use of radiance domain for burn scar detection under varying atmospheric illumination conditions and viewing geometry. Signal, Image and Video Processing. 2017 May 1;**11**(4):605-612

[12] Nguyen D, Ayhan B, Kwan C. A comparative study of several supervised target detection algorithms for hyperspectral images. In: IEEE Ubiquitous Computing Electronics & Mobile Communication Conference, New York City: IEEE; October 2017

[13] Zhou J, Kwan C, Ayhan B. Improved target detection for hyperspectral images using hybrid in-scene calibration. Journal of Applied Remote Sensing. 2017 Aug;**11**(3):035010

[14] Zhou J, Ayhan B, Kwan C, Eismann M. New and fast algorithms for anomaly and change detection in hyperspectral images. In: International Symposium on Spectral Sensing Research. Springfield: Missouri State University; 2010 Jul

[15] Kwon H, Nasrabadi NM. Kernel RX-algorithm: A nonlinear anomaly detector for hyperspectral imagery. IEEE Transactions on Geoscience and Remote Sensing. 2005 Feb;**43**(2): 388-397

[16] Wang W, Li S, Qi H, Ayhan B, Kwan C, Vance S. Identify anomaly component by sparsity and low rank. In: Proceedings of IEEE WHISPERS. Tokyo: IEEE; 2015 Jun. p. 1-4

[17] Li S, Wang W, Qi H, Ayhan B, Kwan C, Vance S. Low-rank tensor decomposition based anomaly detection for hyperspectral imagery. In: 2015 IEEE International Conference on Image Processing (ICIP). Quebec City: IEEE; 2015 Sep 27. p. 4525-4529

[18] Qu Y, Guo R, Wang W, Qi H, Ayhan B, Kwan C, Vance S. Anomaly detection in hyperspectral images through spectral unmixing and low rank decomposition. In: 2016 IEEE International, Geoscience and Remote Sensing Symposium (IGARSS). Beijing: IEEE; 2016 Jul 10. p. 1855-1858

[19] Qu Y, Qi H, Ayhan B, Kwan C, Kidd R. Does multispectral/hyperspectral pansharpening improve the performance of anomaly detection? In: 2017 IEEE International, Geoscience and Remote Sensing Symposium (IGARSS). Fort Worth: IEEE; 2017 Jul. p. 6130-6133

[20] Ayhan B, Kwan C, Li X, Trang A. Airborne detection of land mines using mid-wave infrared (MWIR) and laser-illuminated-near infrared images with the RXD hyperspectral anomaly detection method. Fourth International Workshop on Pattern Recognition in Remote Sensing. Hong Kong: International Association of Pattern Recognition; 2006 Aug 26

[21] He L, Luo J, Qi H, Kwan C. A comparative study of several unsupervised endmember extraction algorithms to anomaly detection in hyperspectral images. In: International Symposium on Spectral Sensing Research. Missouri: Missouri State University; 2010 Jul

[22] Zhou J, Kwan C. Fast anomaly detection algorithms for hyperspectral images. Journal of Multidisciplinary Engineering Science and Technology. 2015 Sep;**2**(9):2521-2525

[23] Dao M, Kwan C, Ayhan B, Tran TD. Bum scar detection using cloudy MODIS images via low-rank and sparsity-based models. In: IEEE Global Conference on 2016 Signal and Information Processing (GlobalSIP). Washington DC: IEEE; 2016 Dec 7. p. 177-181

[24] Dao M, Kwan C, Koperski K, Marchisio G. A joint sparsity approach to tunnel activity monitoring using high resolution satellite images. In: IEEE Ubiquitous Computing Electronics & Mobile Communication Conference. New York City: IEEE; 2017

[25] Perez D, Banerjee D, Kwan C, Dao M, Shen Y, Koperski K, Marchisio G, Li J. Deep learning for effective detection of excavated soil related to illegal tunnel activities. In: IEEE Ubiquitous Computing, Electronics & Mobile Communication Conference. New York City: IEEE; 2017

[26] Ayhan B, Dao M, Kwan C, Chen HM, Bell JF, Kidd R. A novel utilization of image registration techniques to process mastcam images in Mars rover with applications to image fusion, pixel clustering, and anomaly detection. IEEE Journal of Selected Topics in Applied Earth Observations and Remote Sensing. IEEE. 2017 Jul 19; (99). pp. 1-12

[27] Dao M, Kwan C, Ayhan B, Bell JF. Enhancing Mastcam images for Mars rover mission. In: International Symposium on Neural Networks. Cham: Springer; 2017 Jun 21. p. 197-206

[28] Loncan L, de Almeida LB, Bioucas-Dias JM, Briottet X, Chanussot J, Dobigeon N, Fabre S, Liao W, Licciardi GA, Simoes M, Tourneret JY. Hyperspectral pansharpening: A review. IEEE Geoscience and Remote Sensing Magazine. 2015 Sep;**3**(3):27-46

[29] Chavez P, Sides SC, Anderson JA. Comparison of three different methods to merge multiresolution and multispectral data – Landsat TM and SPOT panchromatic. Photogrammetric Engineering and Remote Sensing. 1991 Mar;**57**(3):295-303

[30] Laben CA, Brower BV, inventors; Eastman Kodak Company, assignee. Process for enhancing the spatial resolution of multispectral imagery using pan-sharpening. United States patent US 6,011,875. 2000 Jan 4

[31] Aiazzi B, Baronti S, Selva M. Improving component substitution pansharpening through multivariate regression of MS + Pan data. IEEE Transactions on Geoscience and Remote Sensing. 2007 Oct;**45**(10):3230-3239

[32] Aiazzi B, Alparone L, Baronti S, Garzelli A, Selva M. MTF-tailored multiscale fusion of high-resolution MS and Pan imagery. Photogrammetric Engineering & Remote Sensing. 2006 May 1;**72**(5):591-596

[33] Vivone G, Restaino R, Dalla Mura M, Licciardi G, Chanussot J. Contrast and error-based fusion schemes for multispectral image pansharpening. IEEE Geoscience and Remote Sensing Letters. 2014 May;**11**(5):930-934

[34] Simões M, Bioucas-Dias J, Almeida LB, Chanussot J. A convex formulation for hyperspectral image superresolution via subspace-based regularization. IEEE Transactions on Geoscience and Remote Sensing. 2015 Jun;**53**(6):3373-3388

[35] Simoes M, Bioucas-Dias J, Almeida LB, Chanussot J. Hyperspectral image superresolution: An edge-preserving convex formulation. In: 2014 IEEE International Conference on Image Processing (ICIP), 2014 Oct 27 pp. 4166-4170. Paris: IEEE

[36] Liu JG. Smoothing filter-based intensity modulation: A spectral preserve image fusion technique for improving spatial details. International Journal of Remote Sensing. 2000 Jan 1;**21**(18):3461-3472

[37] Liao W, Huang X, Van Coillie F, Gautama S, Pižurica A, Philips W, Liu H, Zhu T, Shimoni M, Moser G, Tuia D. Processing of multiresolution thermal hyperspectral and digital color data: Outcome of the 2014 IEEE GRSS data fusion contest. IEEE Journal of Selected Topics in Applied Earth Observations and Remote Sensing. 2015 Jun;**8**(6):2984-2996

[38] Hardie RC, Eismann MT, Wilson GL. MAP estimation for hyperspectral image resolution enhancement using an auxiliary sensor. IEEE Transactions on Image Processing. 2004 Sep;**13**(9):1174-1184

[39] Wei Q, Bioucas-Dias J, Dobigeon N, Tourneret JY. Hyperspectral and multispectral image fusion based on a sparse representation. IEEE Transactions on Geoscience and Remote Sensing. 2015 Jul;**53**(7):3658-3668

[40] Yokoya N, Yairi T, Iwasaki A. Coupled nonnegative matrix factorization unmixing for hyperspectral and multispectral data fusion. IEEE Transactions on Geoscience and Remote Sensing. 2012 Feb;**50**(2):528-537

[41] Licciardi GA, Khan MM, Chanussot J, Montanvert A, Condat L, Jutten C. Fusion of hyperspectral and panchromatic images using multiresolution analysis and nonlinear PCA band reduction. EURASIP Journal on Advances in Signal processing. 2012 Dec 1;**2012**(1):207

[42] Huang W, Xiao L, Wei Z, Liu H, Tang S. A new pan-sharpening method with deep neural networks. IEEE Geoscience and Remote Sensing Letters. 2015 May;**12**(5):1037-1041

[43] Masi G, Cozzolino D, Verdoliva L, Scarpa G. Pansharpening by convolutional neural networks. Remote Sensing. 2016 Jul 14;**8**(7):594

[44] Qu Y, Qi H, and Kwan C. Deep learning based pansharpening algorithm for hyperspectral and multispectral images. Submitted to Computer Vision and Pattern Recognition (CVPR) Conference. 2017

[45] Yan Q, Xu Y, Yang X, Nguyen TQ. Single image superresolution based on gradient profile sharpness. IEEE Transactions on Image Processing. 2015 Oct;**24**(10):3187-3202

[46] Keys R. Cubic convolution interpolation for digital image processing. IEEE Transactions on Acoustics, Speech, and Signal Processing. 1981 Dec;**29**(6):1153-1160

[47] Chan SH, Wang X, Elgendy OA. Plug-and-Play ADMM for image restoration: Fixed-point convergence and applications. IEEE Transactions on Computational Imaging. 2017 Mar;**3**(1):84-98

[48] Zhou J, Kwan C, Budavari B. Hyperspectral image super-resolution: A hybrid color mapping approach. Journal of Applied Remote Sensing. 2016 Jul 1;**10**(3):035024

[49] Kwan C, Budavari B, Bovik AC, Marchisio G. Blind quality assessment of fused World-View-3 images by using the combinations of pansharpening and hypersharpening paradigms. IEEE Geoscience and Remote Sensing Letters. 2017 Aug;**24**

[50] Kwan C, Budavari B, Gao F. A hybrid color mapping approach to fusing MODIS and Landsat images for forward prediction. Submitted to MDPI Journal of Remote Sensing. 2017

[51] Kwan C, Budavari B, Dao M, Ayhan B, Bell JF. Pansharpening of Mastcam images. In: 2017 IEE International, Geoscience and Remote Sensing Symposium (IGARSS). Fort Worth: IEEE; 2017 Jul. p. 5117-5120

[52] Kwan C, Ayhan B, Budavari B. Fusion of THEMIS and TES for accurate Mars surface characterization. In: Proceedings of IEEE International Geoscience and Remote Sensing Symposium. Fort Worth: 2017. p. 3381-3384

[53] Kwan C, Choi JH, Chan S, Zhou J, Budavari B. Resolution enhancement for hyperspectral images: A super-resolution and fusion approach. In: 2017 IEEE International Conference on Acoustics, Speech and Signal Processing (ICASSP). New Orleans: IEEE; 2017 Mar 5. p. 6180-6184

[54] Kwan C, Budavari B, Dao M, Zhou J. New sparsity based pansharpening algorithm for hyperspectral images. In: IEEE Ubiquitous Computing, Electronics & Mobile Communication Conference. New York City: IEEE; 2017

[55] Tropp JA. Greed is good: Algorithmic results for sparse approximation. IEEE Transactions on Information theory. 2004 Oct;**50**(10):2231-2242

[56] Yang J, Zhang Y. Alternating direction algorithms for l-1-problems in compressive sensing. SIAM Journal on Scientific Computing. 2011 Feb 3;**33**(1):250-278

[57] Bayer BE, inventor; Eastman Kodak Company, assignee Color imaging array. United States patent US 3,971,065. 1976 Jul 20

[58] Losson O, Macaire L, Yang Y. Comparison of color demosaicing methods. Advances in Imaging and Electron Physics. 2010 Dec 31;**162**:173-265

[59] Compton JT, Hamilton Jr JF, DeWeese TE, inventors; OmniVision Technologies, Inc., assignee. Image sensor with improved light sensitivity. United States patent US 8,194,296. 2012 Jun 5

[60] Kwan C, Chou B, Kwan LM, Budavari B. Debayering RGBW color filter arrays: A pansharpening approach. In: IEEE Ubiquitous Computing, Electronics & Mobile Communication Conference. New York City: IEEE; 2017

[61] Möller M, Wittman T, Bertozzi AL, Burger M. A variational approach for sharpening high dimensional images. SIAM Journal on Imaging Sciences. 2012 Jan 24;**5**(1):150-178

[62] Zhou J, Kwan C, Ayhan B, Eismann MT. A novel cluster kernel RX algorithm for anomaly and change detection using hyperspectral images. IEEE Transactions on Geoscience and Remote Sensing. 2016 Nov;**54**(11):6497-6504

Chapter 9

Thresholding Algorithm Optimization for Change Detection to Satellite Imagery

René Vázquez-Jiménez, Rocío N. Ramos-Bernal,
Raúl Romero-Calcerrada, Patricia Arrogante-Funes,
Sulpicio Sanchez Tizapa and Carlos J. Novillo

Additional information is available at the end of the chapter

http://dx.doi.org/10.5772/intechopen.71002

Abstract

To detect changes in satellite imagery, a supervised change detection technique was applied to Landsat images from an area in the south of México. At first, the linear regression (LR) method using the first principal component (1-PC) data, the Chi-square transformation (CST) method using first three principal component (PC-3), and tasseled cap (TC) images were applied to obtain the continuous images of change. Then, the threshold was defined by statistical parameters, and histogram secant techniques to categorize as change or unchanged the pixels. A threshold optimization iterative algorithm is proposed, based on the ground truth data and assessing the accuracy of a range of threshold values through the corresponding Kappa coefficient of concordance. Finally, to evaluate the change detection accuracy of conventional methods and the threshold optimization algorithm, 90 polygons (15,543 pixels) were sampled, categorized as real change/unchanged zones, and defined as ground truth, from the interpretation of color aerial photo slides aided by the land cover maps to obtain the omission/commission errors and the Kappa coefficient of agreement. The results show that the threshold optimization is a suitable approach that can be applied for change detection analysis.

Keywords: Landsat, threshold, histogram, change detection, optimization, algorithm

1. Introduction

The change detection is the process of identifying differences between the state of specific characteristics of a phenomenon or elements, by comparing its spatial representation at two points in time [1–3], controlling all variances caused by the differences in variables that are not of interest, and measuring changes caused by differences in the variables of interest.

The change detection applied to assess the transformation of the space over the time can be a suitable alternative to support the territory management policies and in the proposal of appropriate actions of urban planning, natural resource management, disaster assessment, and risk management that are required to cope with the impacts of global change that are recorded in several regions of the world and are rapidly transforming the landscapes [3–6].

To evaluate and address suitably the effect of changes in land coverage and land use, several studies and models of change detection based on multispectral image data analysis have been carried out such as principal component analysis (PCA), image subtraction, image ratioing, change vector analysis, image regression, and spectral features variance, which are the usual methods of the change detection assessment [7–11].

Other nonconventional studies introduce transformations as input data in change detection to process, such as the use of the pseudo-cross variogram as a statistical function to quantify the changes [4], the multivariate alteration detection, based on the established canonical correlations analysis, introduced by [12], or the Chi-square transformation, based on the Mahalanobis distance between pixels of images at different dates, submitted by Ridd and Liu [13].

The first stage of any change detection process results in a (continuous) change image, in which the pixel value corresponds to the intensity of the change calculated according to the method employed. However, to complete the process of detecting changes, it is necessary the generation of a binary map that categorizes the continuous value of the pixels only into two classes: change and no change. Thus, a segmentation of the continuous change image is required, according to a value assigned as a threshold, and thus, the dynamic areas can be distinguished from those that remained stable during the period analyzed [10, 14].

Considering that only a few pixels change over a period, the classic method for defining the threshold value needed to categorize the changed and unchanged pixels is based on the fact that the density function of the difference image can be considered almost equal to the density function of unmodified pixels [15]. Thus, the unchanged pixels are distributed around the mean (μ), while the changes are spread over the tails of the distribution [1–16], separated from the mean a certain number of standard deviations of the distribution.

Typically, the definition of the threshold values is directly based on a manual process of trial-and-error or by using the statistical parameters (mean and standard deviation) of the difference image interactively, adjusting the number of standard deviations and evaluating the results until the sets of accuracy and error are satisfied [7, 10, 17]. In practice, the determination of the thresholds is highly subjective, based on the analyst's abilities and the known details of the study area [18].

There are also other studies where automatic processes define the threshold value. Hervás and Rosin [19] proposed an automatic threshold definition through the intensity value of the pixel in the difference image, which corresponds to the point in the histogram (usually unimodal) of maximum distance to the secant line between the highest and lowest points of the distribution function [20].

In the first stage of this work, two conventional methods of detection of changes are applied: linear regression (LR) and Chi-square transformation (CST); followed by the thresholding definition by statistical parameters and by the histogram secant technique, in order to categorize the pixels as changes or no change and thus generate the final thematic map for each method.

In a second stage, an optimization algorithm was designed, aimed to define the value of an optimal threshold that maximizes the pixel hits and minimizes the fails ones, and based on the ground truth data of the studied area. This stage was developed for each of the continuous images obtained from the conventional methods analyzed, and also, a final (optimal) thematic change map was obtained.

Finally, the accuracy of each of the final thematic maps obtained from the traditional methods and the threshold optimization process was assessed, validating the results with the collected ground truth information through confusion matrices.

The Dinamica EGO software[1] [21, 22] was used to develop the described models and processes of this work.

2. Materials and methods

2.1. Study area

The studied area consists of a group of basins in the central region of the Guerrero state in México, around the largest gold mine in Latin America "Filos-Bermejal" [23] covering approximately 4800 km^2 (**Figure 1**), of which 43.8% is covered by oak, pine, and mesophyll forest, 39.3% is deciduous forest, 8.2% is agricultural-livestock land use, and 8.7% is covered by grassland, bare soil, water, and urban.

This area is interesting because the concentration of diverse factors that lead to an important dynamics of landscape transformation, such as the pressure on vegetation areas by urban and agricultural growth, there are deforestation areas by industrial and mining activity and also areas of natural hazards.

2.2. Dataset

Two Landsat surface reflectance (SR) images (Path 26, Row 48, WRS-2) were used. This processed product is courtesy of the U.S. Geological Survey Earth Resources Observation and Science Centre and is offered as climate data records [24, 25]. The images correspond to the dates and sensors: February 24, 2011 for Landsat 5-TM and February 16, 2014 for Landsat 8-OLI.

[1]Dinamica EGO consists of a sophisticated platform for environmental modeling with outstanding possibilities for the design from the very simple static spatial model to very complex dynamic ones, which can ultimately involve nested iterations, multi-transitions, dynamic feedbacks, multi-region and multi-scale approach, decision processes for bifurcating and joining execution pipelines, and a series of complex spatial algorithms for the analysis and simulation of space-time phenomena [22].

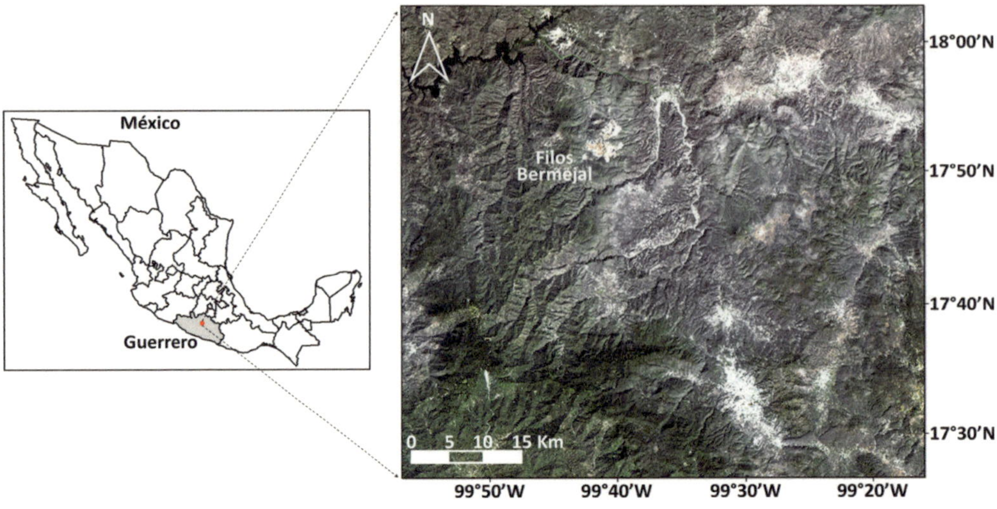

Figure 1. Mining zone "Filos-Bermejal" in Guerrero State, México.

In the performed processes, the thermal bands were excluded. Thus, the bands used were Blue, Green, Red, Near-Infrared (NIR), Shortwave Infrared 1 (SWIR-1), and Shortwave Infrared 2 (SWIR-2), according to the official nomenclature [26].

The SR data are produced by the specialized software *Landsat Ecosystem Disturbance Adaptive Processing System* for Landsat 5-TM and *L8SR* for Landsat 8-OLI, developed by the Making Earth System Data Records for Use in Research Environments of the Goddard Space Flight Center of NASA and the University of Maryland [27].

The software applies the MODIS atmospheric correction routines. In addition to the Landsat data, the water vapor, ozone, geopotential height, aerosol optical thickness, and digital elevation data are presented to the radiative transfer model *Second Simulation of a Satellite Signal in the Solar Spectrum* (6S) to generate the top of atmosphere reflectance, SR, brightness temperature and cloud masks, cloud shadows, and adjacent clouds, land, and water.

2.3. Topographic correction and image pre-processing

An overview of the stages performed in this study can be seen in **Figure 2**.

To prevent false results in the change detection analysis, it is recommended to perform a precise coregistration, radiometric calibration, atmospheric, and topographic corrections between the multitemporal images [28, 29]. Almost all of these processes are already applied in the elaborated product used (Landsat SR images).

Thus, to reduce the effect that the changes of slopes, terrain orientation, and solar geometry at the time of the image acquisition cause on the reflectance values, the topographic correction was performed by Sun Canopy Sensor + Correction (SCS + C) method [30], which is more

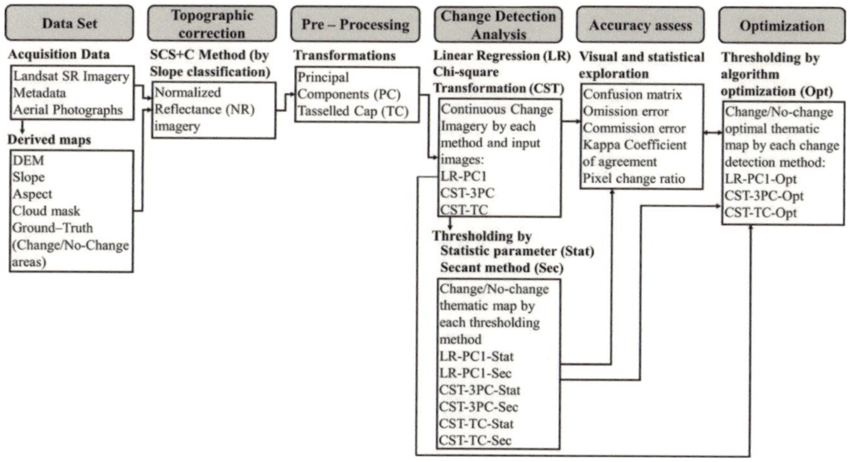

Figure 2. Methodology overview.

appropriate for forest mountain areas than other ground-based methods, because it preserves the geotropic nature of the trees (growth normal to geoid) [31].

The C parameter used in the topographic correction aims to improve the correction by moderating the overcorrection of dimly lit pixels [32] and was determined by linear regressions between the illumination and reflectance values, according to a classification of the topographical slopes of the studied area [14, 33]. Thus, six NR bands for each image date were obtained. A true-color composite of the normalized bands (RGB) is shown in **Figure 3**.

Images of brightness, greenness, and wetness indices (tasseled cap-TC) [34–36] and principal components (PC) [37] were generated, to be processed as input data in the change detection analysis to differentiate the results.

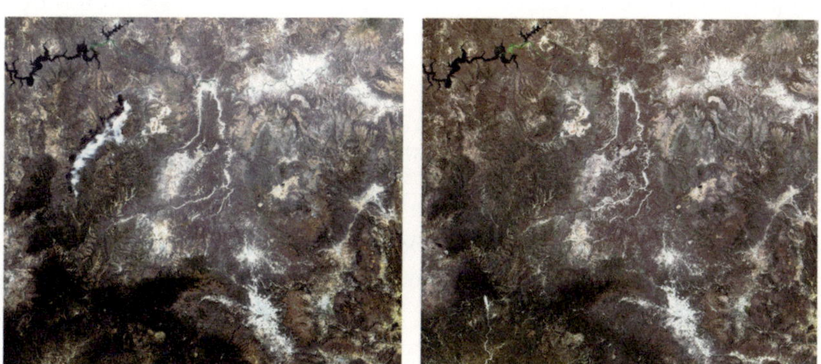

Figure 3. February 24, 2011 for Landsat 5-TM (left) and February 16, 2014 Landsat 8-OLI, true-color images topographically corrected.

2.4. Change detection methods

2.4.1. Linear regression (LR)

The LR method for change detection assumes that the pixel values (Y) of the final date image f_2 results from a linear function of the pixel values (X) of the initial date image f_1. Thus, it is possible to perform a regression from $Y^k_{i,j}(f_2)$ to $X^k_{i,j}(f_1)$ by least squares [1, 13, 38] to obtain the slope m and ordinate b regression line parameters to find an equation of the form $Y = mX + b$ to model it.

Applying the RL model to the f_1 image, a new image $Y'^k_{i,j}(f_1)$ can be built, which corresponds to the expected values generated by the prediction model. Through the expected and actual values of f_2 image, it is possible to obtain a continuous image of the residual values calculated by:

$$R^k_{i,j} = Y'^k_{i,j}(f_1) - Y^k_{i,j}(f_2) \tag{1}$$

where $R^k_{i,j}$ is the residual pixel value of i line and j column for the k band.

If no changes are recorded (Residual = 0), then the pixel values obtained by the prediction model in the expected image will be the same as that of the real pixel values in the final date image f_2. On the other hand, the founded differences registered in the corresponding residual value image will indicate a change, and its magnitude will indicate the intensity.

The first PC images (PC-1) of each date were used as input data in the performance of the LR detection method, obtaining a single image of continuous change.

2.4.2. Chi-square transformation (CST)

The CST is a statistical technique that is applied to obtain a measure of divergence or distance between groups regarding multiple characteristics. The most used measure is *Mahalanobis distance* (Md) [39] and plays a fundamental role in the data analysis with multiple measurements, finding applications on statistical patterns recognition in archeology, medical diagnosis, or remote sensing [40].

In remote sensing, the Chi-square transformation is applied to recognize patterns in a single global image, starting from the multivariate information stored in numerous multispectral bands [20].

To distinguish how much each image has changed over the time, an additional image of the relative difference of each pixel (P) compared to its mean (μ) was calculated by $(P - \mu)/\mu$, applied to each kind of images and dates [20]. For this task, a cloud mask was considered in the calculation of the mean not to affect the final values of relative differences.

For each pair of corresponding images acquired on different dates, a simple subtraction pixel-by-pixel was performed, aimed at obtaining a third image that stores the numerical differences. Thus, three differences images have been achieved for each kind of data: Dif-PC_1, Dif-PC_2, and Dif-PC_3 and Dif-TC_B, Dif-TC_G, and Dif-TC_W.

http://dx.doi.org/10.5772/intechopen.71002

According to Ridd and Liu [13], the CST detection method was applied to each group of difference images (Dif-PC and Dif-TC) to obtain a single global image that stores the continuous change, represented by the square Md [39, 40], determined by the equation:

$$Y = (X - M)^T \sum^{-1} (X - M) \tag{2}$$

where Y is the square root of Md for each pixel in the change image, X is the difference vector of the n values between the two dates, M is the vector of the mean residuals of each band, T is the transpose of the matrix $(X - M)$, and $(P - 1)$ is the inverse covariance matrix of the bands between the two dates.

The Md takes into account the variance of each variable and the covariance between variables. Geometrically, this is done by transforming the data into standardized uncorrelated data and computing the ordinary Euclidean distance for the transformed data, thus providing a way to measure distances that take into account the scale of the data (standard deviation) [20, 41].

The Md performs a Chi-square distribution with degrees of freedom equal to the number of bands used in the transformation model. Thus, it is possible to compare the distances just as the data would be compared to a normal distribution, when the number of bands has a multivariate normal distribution with the covariance matrix Σ and the vector of the mean residuals M [13, 20, 42].

The square root of the Md (Y) can be represented as a single image, and the pixel values represent the magnitudes of change between the analyzed dates. Thus, theoretically, a zero pixel value means an absolute no change.

The CST was applied to the first 3-PC images and the TC images between the dates of study. Thus, two single continuous images of change were obtained.

2.5. Threshold definition

Any of the methods explained previously results in a continuous change image, where the pixel values correspond to the intensity of the change estimated according to the method employed. However, to complete the process, it is necessary to generate a categorical binary map that isolates the pixels into change and no change classes. Thus, it is a need to segment the continuous change image according to a value assigned as a threshold, and in this way, the dynamic areas can be distinguished from those that were stable during the analyzed period.

2.5.1. Statistical method

From the continuous change image, the statistical parameters of the mean (μ) and the standard deviation (σ) are used to calculate the threshold by $\mu \pm n\sigma$.

According to D'Addabbo et al. [16], the density function of the continuous change image is almost equal to the density function of the unmodified pixels, and in the determination of the threshold statistically fixed, n is an empirical parameter set by the user that can be adjusted.

In this stage, the n value was set equal to 2. Thus, the $n\sigma$ factor was obtained first, then added to, and subtracted from the mean value of a continuous image obtained by the LR method (with thresholds in both tails of the distribution, as can be seen in **Figure 3**). On the other hand, the $n\sigma$ factor was only added to the average value of the continuous image obtained by the CST method (with a single threshold in the right tail of the distribution).

2.5.2. Secant method

The threshold is automatically defined by selecting the pixel value corresponding to the point in the distribution of the histogram, where the maximum perpendicular line intersects the secant line between the highest and lowest points of the histogram [19]. According to the distribution of the continuous change image, this procedure was performed just in the right tail for the CST detection method and in both tails for the LR detection method (**Figure 4**).

2.6. Accuracy assessment

To assess the accuracy of the change detection process, a sampling of 90 polygons was identified as ground truth, and their pixels were categorized as change and unchanged. Through the interpretation of color, aerial photo slides near to the study dates, aided by the land cover, land use, and vegetation maps, 7810 pixels were defined as changed and 7733 ones were defined as unchanged. This ground truth was compared with the final thematic maps obtained from each detection method, and each thresholding technique applied through confusion matrices to get the omission and commission errors.

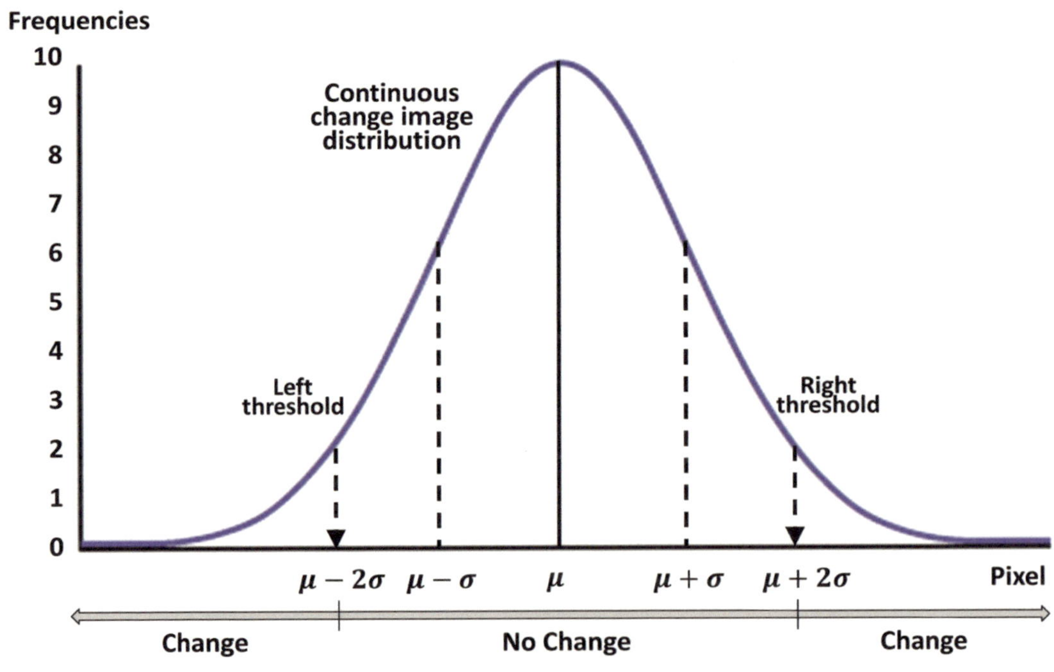

Figure 4. Scheme of the definition of thresholds by the statistical method in a normal distribution.

For each final thematic map, the Kappa concordance coefficient of agreement [43] was obtained to quantify the difference between the observed map-reality agreement and the one that would be expected simply by random. The Kappa index attempts to define the degree of adjustment due only to the accuracy of the categorization, regardless of random factors causes [10, 44]. The Kappa coefficient was calculated by:

$$k = \left(n \sum_{i=1,n} X_{ii} - \sum_{i=1,n} X_{i+} X_{+i}\right) / \left(n^2 - \sum_{i=1,n} X_{i+} X_{+i}\right) \quad (3)$$

where k is the Kappa coefficient of agreement, n is the sample size, X_{ii} is the observed agreement, and $X_{i+} X_{+i}$ is the expected agreement in each category i.

The Kappa coefficient allows knowing if the marked degree of agreement draws away or is not significantly different from the expected random agreement. The agreement observed highlights on the diagonal of the confusion matrix, and the expected agreement is used to calculate the fit between the map and the reality, due to randomness [10].

2.7. Threshold optimization

The optimization algorithm proposed is based on the Kappa concordance coefficient of agreement, considering the threshold value that reports the highest Kappa index corresponds to the one that also will report the maximum numbers of pixels hit and the minimum numbers of pixels fail.

The algorithm is an automatic, but also a supervised process, unlike previous methods explained that are automatic unsupervised. Thus, it is necessary to have real information as field data about the changed and unchanged areas that will be used as ground truth (the same data to assess the accuracy) to train and evaluate the optimization iteratively.

It is also necessary for an initial threshold value, to start the iterative process, which can be any of the automatically threshold determined by the previously explained methods (statistical or secant).

Figure 6 shows the flowchart of the optimization process.

For the CST detection method (using 3-PC and TC as the input images), the threshold optimization process was applied following the scheme as it is shown in **Figure 5**; starting from the value of the threshold established by the statistical method to set the range and develop 200 iterations in the whole process.

For the continuous change image obtained by the LR detection method (using PC-1 as the input image), the threshold optimization process was applied in two stages, first looking for the optimal value of the right threshold by establishing the left one equal to the value calculated by the statistical method. Then, the optimum value of the left threshold was sought by setting as the right limit equal to the optimal value previously founded.

For each method, two maps (change/no change) were obtained, and their accuracy was assessed by confusion matrices and the concordance index (Kappa). Then, the proposed algorithm was performed to obtain an optimal threshold by successive iterations to first increasing/

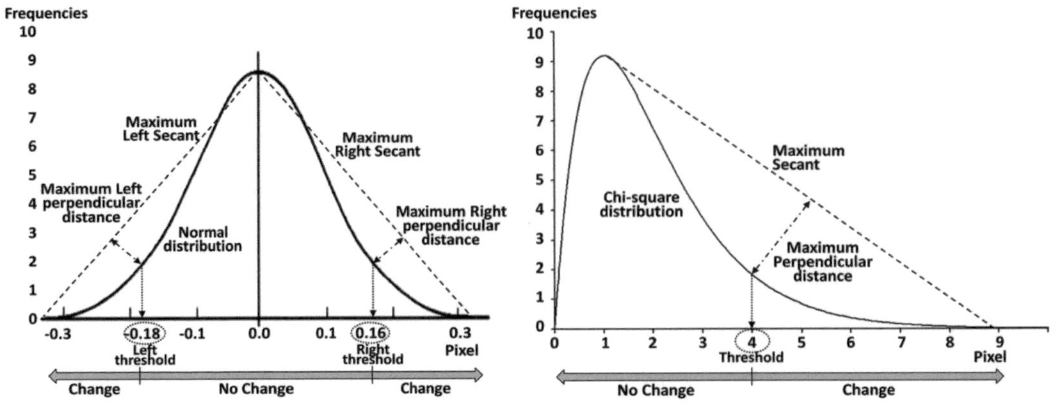

Figure 5. Scheme of the definition of thresholds by the secant method. Left on a normal distribution. Right on a Chi-square distribution.

decreasing the last threshold value and then calculate the corresponding Kappa and comparing with previous results to stop the process and finally define the highest Kappa linked to the optimal threshold.

For each continuous change, image generated by each detection method and kind of input image processed, an optimal final binary change/no change map was obtained, which was compared, analyzed, and discussed.

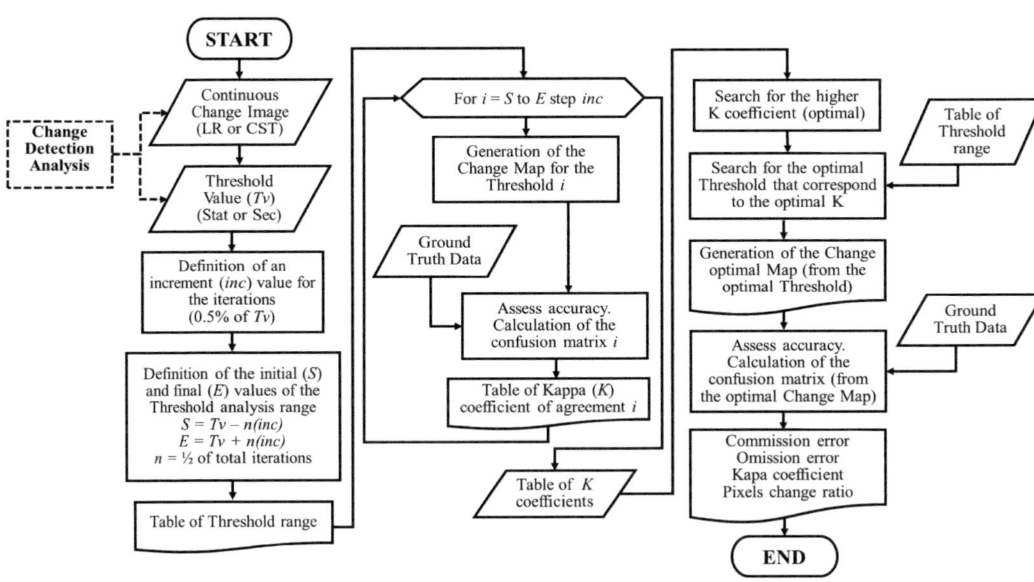

Figure 6. Flowchart of the threshold optimization algorithm.

3. Results and discussion

3.1. Change detection analysis

The threshold definition processes performed to the histograms of the continuous change image arising from the LR detection method using PC-1 as input images are shown in **Figure 7**. The values raised from the thresholding (statistical and secant) processes are also illustrated.

The observed ranges of the distribution (**Figure 7**) show the changed pixels (at the ends of both tails) for each of the threshold methods applied and also the unchanged pixels remaining in the center of the distribution (around the mean).

The definition of the thresholds values varies according to the applied method. The threshold values in the left tail are similar in both approaches; however, in the right tail, the statistical method reports 0.0818 as a threshold, while the secant method reports 0.0693. This fact affects the number of pixels considered as changed and their location in the final thematic change maps, making it more or less intense visually.

The histograms of the continuous change images resulting from the CST detection method using the 3-PC and TC as input images, as well as the values arising from the thresholding process are shown in Appendix A.

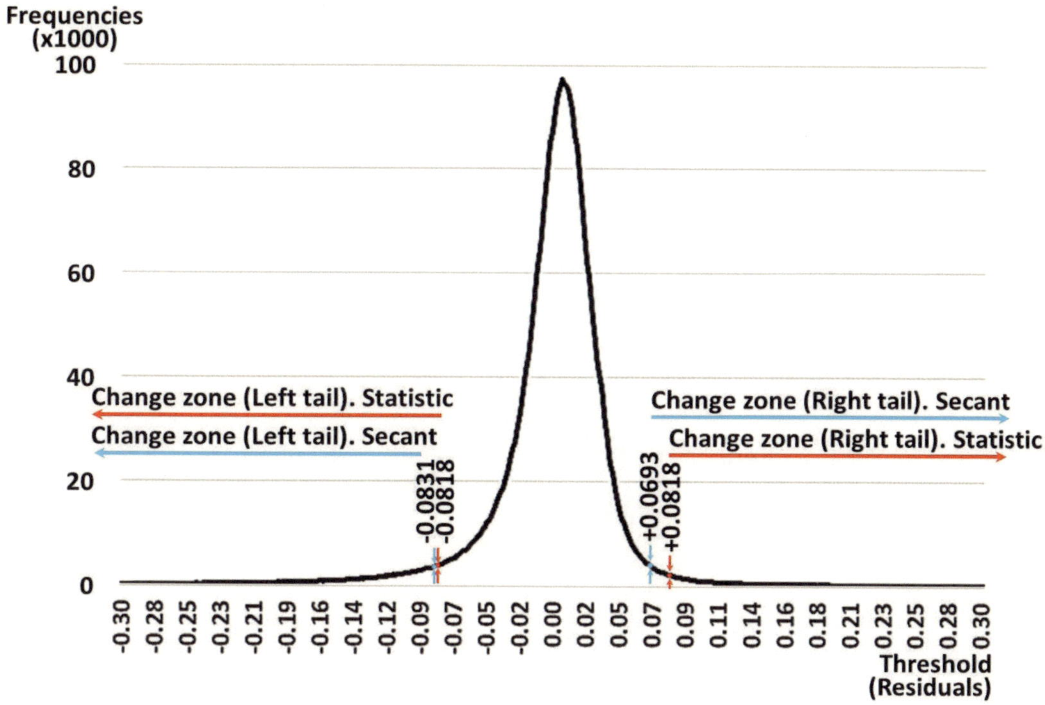

Figure 7. Threshold values obtained by statistical and secant methods on the distribution of the continuous change image from the LR detection method of PC-1 images.

3.2. Optimization process

The optimization algorithm was applied to each of the three detection methods considered in the study. **Figure 8** shows the values of the coefficient of concordance Kappa reached by each threshold valued analyzed for each of the detection method and threshold definition.

The algorithm determines the Kappa index for each of the threshold values within the analyzed range. According to the patterns depicted, it can be seen how the value of the tested threshold increases, the successes increase, and errors decrease, which is reflected in a better Kappa index. This performance occurs until a certain limit (maximum of the curve); from this point, the pattern change, now the successes decrease, the errors increase, and the Kappa index decreases. Then, based on the pattern depicted for each detection method, the algorithm determines the highest Kappa index and matches it to its corresponding threshold, resulting as the optimal one.

3.3. Thematic maps of change

From the threshold values determined for each detection method applied, thematic change maps were generated. **Figure 9** shows the maps obtained by the LR detection method and each thresholding determined.

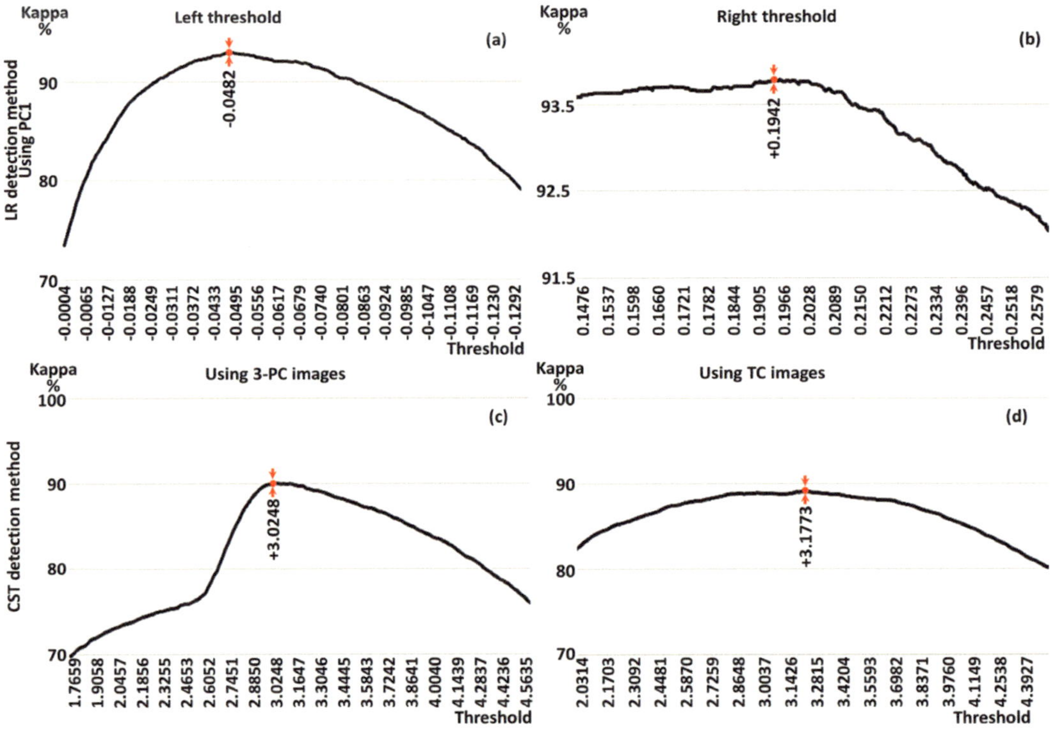

Figure 8. Determination of optimal-threshold values based on Kappa index. (a, b) LR using PC-1, (c) CST using 3-PC, and (d) using TC.

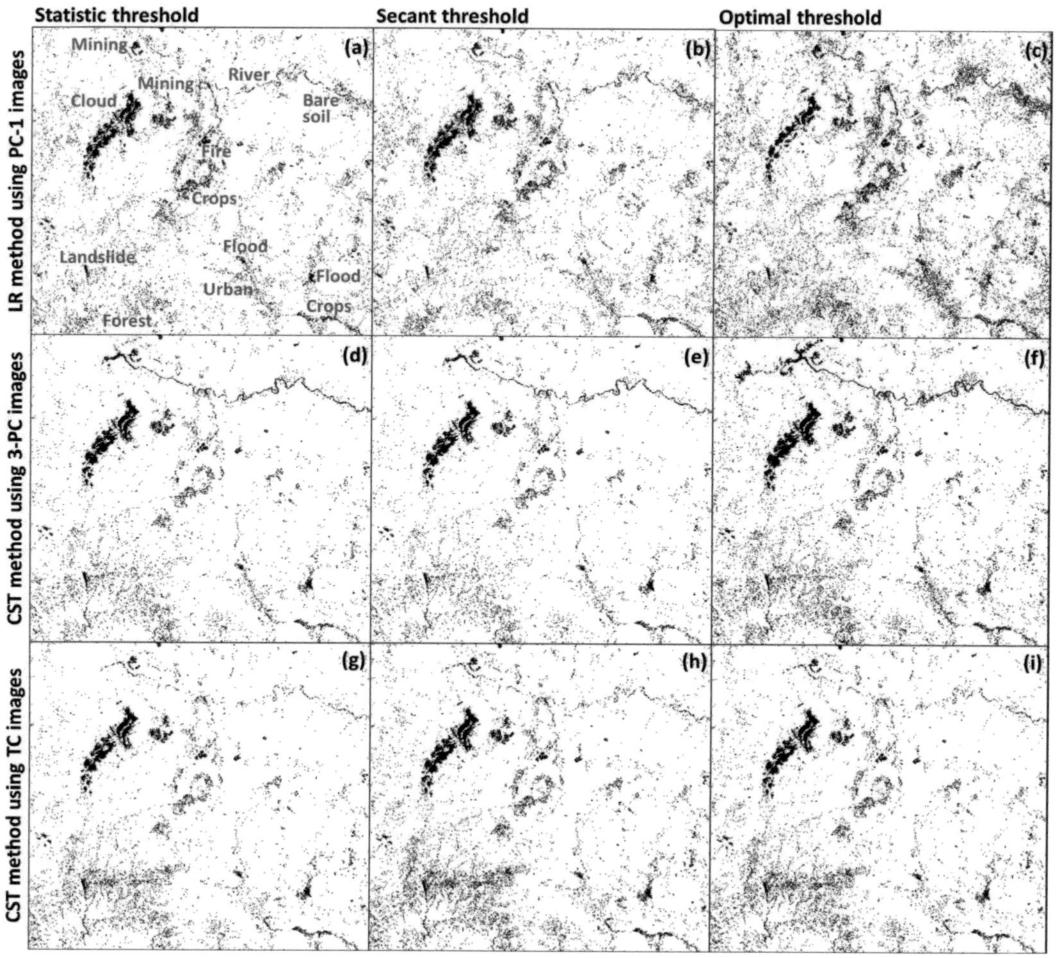

Figure 9. Thematic change maps obtained by LR detection method. (a) Statistic threshold, (b) secant threshold, (c) optimal threshold.

The resulting thematic maps seem similar, detecting changes around the same places regardless of the detection method, the nature of the input images used, and the thresholding technique. At a glance, the resulting maps show the agreement of the analyzed approaches in detecting visible changes caused by specific phenomena or events occurred between the dates (**Figure 9a**).

The resulting maps are more or less visually intense, depending on whether more or fewer pixels are detected as change. According to the kind of data used as input images and how it is analyzed by the detection method, result in more or less homogeneous areas represented in maps for certain kinds of land cover use.

More pixels can be detected as a change in the river zones when 3-PC images were used (**Figure 9(d–f)**) may be due to the presence of elements (such as sediments in water) that this kind of images detects better than the other used images. In forest areas, more changes are detected when TC images were used (**Figure 9(g–i)**)]. As the previous case may be, the nature of the data is more suitable for detect shifts in this type of cover.

The optimal-threshold map obtained by the LR method seems to detect more pixels as a change in the same areas as the statistic and secant maps, except in the cloud zone where a difference is observed in the interpretation as a change between the cloud and its shadow (according to **Figure 3**).

In the same way, the optimal-threshold map obtained by the CST using 3-PC images seems more visually intense than the statistic and secant maps in all areas, including the cloud zone. While the optimal-threshold map from the CST using TC images looks very similar to the statistical and secant maps.

3.4. Accuracy assessment

To assess the accuracy of the resulting thematic change maps, confusion matrices were performed to obtain the omission and commission errors and the Kappa coefficient of agreement (**Table 1**).

According to the ground truth validation, all the analyzed methods reach at least 85% of accuracy in isolating the changed from the unchanged areas. According to the Kappa coefficient of agreement (**Table 1**), it can be seen that the LR method results in an improvement in the change detection process in comparison to the CST method.

The CST method reports very similar Kappa coefficients, regarding the use of 3-PC or TC as input images in the detection of general changes. However, as was mentioned before, one kind of images could have better results than another, for the detection of specific changes and specific land covers.

Detection method	Input image	Threshold definition	Threshold value	Pixels detected as change (%)	Mean omission error	Mean commission error	Kappa
LR	PC-1	Statistic	<–0.0818 > 0.0818	4.40	4.80	4.70	90.31
LR	PC-1	Secant	<–0.0831 >0.0693	4.89	5.00	4.90	90.07
LR	PC-1	Optimal	<-0.0482 > 0.1942	7.37	3.10	3.10	93.78
CST	3-PC	Statistic	>3.4969	3.64	6.00	5.80	87.97
CST	3-PC	Secant	>3.5882	3.41	6.30	6.10	87.29
CST	3-PC	Optimal	>3.0248	5.44	5.00	5.00	90.04
CST	TC	Statistic	>3.4725	3.61	5.80	5.80	88.29
CST	TC	Secant	>3.1137	4.76	5.60	5.60	88.75
CST	TC	Optimal	>3.1773	4.53	5.50	5.50	89.00

Table 1. Threshold values, pixel change ratio, omission/commission error, and the kappa coefficient of agreement, obtained from each detection method, input image, and threshold technique.

According to the Kappa coefficient, in all cases, the optimal-threshold thematic change map reports an improvement compared with the accuracy by statistic and secant thresholds. Over the mean of statistic and secant coefficients, the improvement is 3.6% for the LR method, 1.6% for the CST method using 3-CP images, and 0.3% for the CST method using TC images (**Table 1**).

From all the analyzed approaches, the LR detection method using 1-PC as input images and the threshold value determined by the optimization algorithm combine the highest Kappa coefficient and the lowest proportion of omission–commission errors; thus it is the most accurate in isolating the changed from the unchanged areas, according to the ground truth validation defined.

On the other hand, the CST detection method using TC as input images and the threshold value determined by statistic parameters is the least accurate of analyzed approaches.

4. Conclusions

In the change detection analysis, the threshold value designation is a major step that defines the overall capacity of the method employed. A suitable threshold maximizes its ability to discriminate the dynamic from the stable areas; but even other impacts such as those caused by random factors occurring between the periods analyzed could also be detected like differences in illumination, atmospheric conditions, soil moisture, sensor calibration, or information recording.

The statistical threshold definition is suitable when the pixel ratio of change in the studied area is less than ≈4.6%, by the threshold definition based on the statistical parameters. On the other hand, the threshold optimization algorithm works with the information of the sample data of the ground truth and thus has no limitation of the pixel ratio of change resulting, as can be seen in **Table 1**.

According to the obtained results, it can be concluded that the proposed threshold optimization algorithm is a suitable approach in the change detection analysis, since it reports substantial improvements compared with the other conventional detection methods developed.

The proposed threshold optimization is a supervised approach, and as such, it necessarily requires ground truth information to be performed which could be a limitation, unlike the conventional methods applied which are not supervised and fully automatic in the definition of thresholds.

Although this studio focuses on Landsat imagery, the threshold optimization can be applied to any satellite imagery, any study area, and any data information contained in the images to compare.

The threshold optimization could even be of interest in areas other than satellite imagery, such as medical imaging where images of the human body are used in medical procedures that seek to reveal, diagnose, or examine diseases.

Appendix

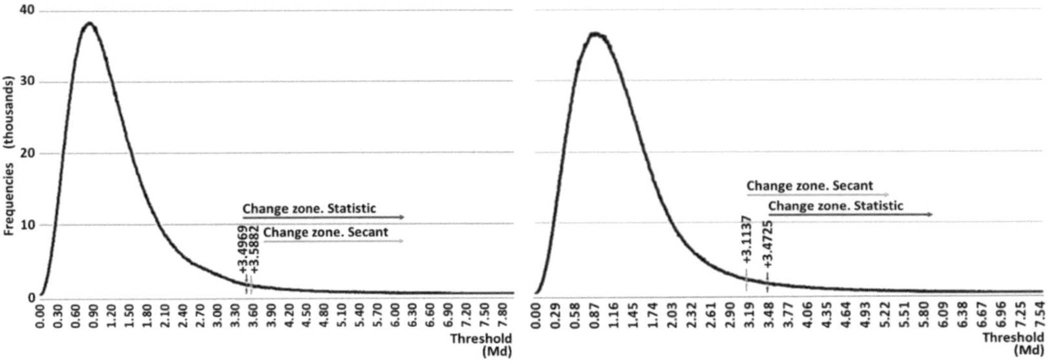

Appendix A. Threshold values obtained by statistical and secant methods on the distribution of the continuous change image from the CST detection method of 3-PC (left) and TC (right) images.

Author details

René Vázquez-Jiménez[1,3]*, Rocío N. Ramos-Bernal[1,3], Raúl Romero-Calcerrada[2], Patricia Arrogante-Funes[3], Sulpicio Sanchez Tizapa[1] and Carlos J. Novillo[3]

*Address all correspondence to: rvazquez@uagro.mx

1 Academic Body UAGro CA-93 Natural Hazards and Geotechnology, Autonomous University of Guerrero, Chilpancingo, Mexico

2 Faculty of Legal and Social Sciences, King Juan Carlos University, Madrid, Spain

3 Department of Chemical and Energy Technology, Chemical and Environmental Technology and Mechanical Technology King Juan Carlos University, Madrid, Spain

References

[1] Singh A. Review article digital change detection techniques using remotely-sensed data. International Journal of Remote Sensing. 1989;**10**(6):989-1003. DOI: 10.1080/014311689 08903939

[2] Wang Z, Bovik AC, Sheikh HR. Fast subpixel mapping algorithms for subpixel resolution change detection. IEEE Transactions on Geoscience and Remote Sensing. 2015;**53**(4):1692-1706. DOI: 10.1109/TGRS.2014.2346535

[3] Zhang P, Gong M, Su L, Liu J, Li Z. Change detection based on deep feature representation and mapping transformation for multi-spatial-resolution remote sensing images. ISPRS Journal of Photogrammetry and Remote Sensing. 2016;**116**:24-41. DOI: 10.1016/j.isprsjprs.2016.02.013

[4] Chica-Olmo M, Ruiz-Castellano J, Rigol-Sánchez J. Aplicación de la función Variograma al Análisis de Cambios Espacio-Temporales en imágenes Landsat TM. Revista de Teledetección. 2004;**22**:41-46

[5] Xiaodong Z, Ya G, Deren L. A strategy of change detection based on remotely sensed imagery and GIS data. In: Norman AS, editor. Proceedings of the ISPRS Commission. VII Symposium "Remote Sensing: From Pixels to Processes". The Netherlands: Enschede; 2006

[6] Zanchetta A, Bitelli G, Karnieli A. Monitoring desertification by remote sensing using the tasselled cap transform for long-term change detection. Natural Hazards. 2016;**83**(1):223-237. DOI: 10.1007/s11069-016-2342-9

[7] Mas J. Monitoring land-cover changes: A comparison of change detection techniques. International Journal of Remote Sensing. 1999;**20**(1):139-152. DOI: 10.1080/014311699213659

[8] Xu L, Zhang S, He Z, Guo Y. The comparative study of three methods of remote sensing image change detection. In: 17th International Conference on Geoinformatics; 12-14 Aug. 2009; Fairfax, VA, USA. IEEE; 2009. p. 1595-1598. DOI: 10.1109/GEOINFORMATICS.2009.5293490

[9] Mas JF, Kolb M, Houet T, Paegelow M, Camacho M. Una comparación de diferentes enfoques de modelación de cambios de cobertura/uso del suelo. In: XIV Simposio Internacional SELPER 2010. Nov. 2010. Guanajuato, México. SELPER; 2010. pp CD, 2010

[10] Chuvieco E. Teledetección Ambiental: La observación de la tierra desde el espacio. 1st ed. Madrid, Spain: Ariel; 2010. 591 p. ISBN 978-84-344-3498-1

[11] Hussain M, Chen D, Cheng A, Wei H, Stanley D. Change detection from remotely sensed images: From pixel-based to object-based approaches. ISPRS Journal of Photogrammetry and Remote Sensing. 2013;**80**:91-106. DOI: 10.1016/j.isprsjprs.2013.03.006

[12] Nielsen AA, Conradsen K, Simpson JJ. Multivariate alteration detection (MAD) and MAF post processing in multispectral, bitemporal image data: New approaches to change detection studies. Remote Sensing of Environment. 1998;**64**(1):1-19. DOI: 10.1016/S0034-4257(97)00162-4

[13] Ridd MK, Liu J. A comparison of four algorithms for change detection in an urban environment. Remote Sensing of Environment. 1998;**63**(2):95-100. DOI: 10.1016/S0034-4257(97)00112-0

[14] Vázquez-Jiménez R. Método Alternativo para la Elaboración de Cartografía Consistente de Ocupación y Uso del Suelo, Orientado a Estudios de Dinámica Territorial [thesis]. Madrid, Spain: Universidad Rey Juan Carlos; 2017. 267 p

[15] Roy M, Ghosh S, Ghosh AA. novel approach for change detection of remotely sensed images using semi-supervised multiple classifier system. Information Sciences. 2014;**269**:35-47. DOI: 10.1016/j.ins.2014.01.037

[16] D'addabbo A, Satalino G, Pasquariello G, Blonda P. Three different unsupervised methods for change detection: An Application. In: 2004 International Geoscience and Remote Sensing Simposium; 20-24 Sept-2004; Anchorage, AK, USA. IEEE; 2004. DOI: 10.1109/IGARSS.2004.1370735

[17] Deilami BR, Ahmad BB, Saffar MR, Umar HZ. Review of change detection techniques from remotely sensed images. Research Journal of Applied Sciences, Engineering and Technology. 2015;**10**(2):221-229. DOI: 10.19026/rjaset.10.2575

[18] Lu D, Li G, Moran E. Current situation and needs of change detection techniques. International Journal of Image and Data Fusion. 2014;**5**(1):13-38. DOI: 10.1080/19479832.2013.868372

[19] Hervás J, Rosin PL. Tratamiento digital de imágenes de Teledeteción en el espectro óptico para el reconocimiento y control de deslizamientos . In: V Simposio Nacional sobre taludes y laderas inestables; 27-30 Nov 2001. Madrid, Spain. Ministerio de Fomento; 2001. p. 63-74

[20] Vázquez-Jiménez R, Romero-Calcerrada R, Novillo CJ, Ramos-Bernal RN, Arrogante-Funes P. Applying the chi-square transformation and automatic secant thresholding to landsat imagery as unsupervised change detection methods. Journal of Applied Remote Sensing. 2017;**11**(1):1-14. DOI: 10.1117/1.JRS.11.016016

[21] Soares-Filho BS, Rodrigues HO, Costa W. Modeling Environmental Dynamics with Dinamica EGO. Vol. 115. Minas Gerais: Centro de Sensoriamento Remoto, Universidade Federal de Minas Gerais, Belo Horizonte; 2009

[22] Centro de Sensoriamento Remoto-Universida de Federal de Minas Gerais. What is Dinamica EGO? [Internet]. 2015. Available from: http://csr.ufmg.br/dinamica/ [Accessed: Aug-23-2017]

[23] Notimex. En Guerrero, la mina más grande de oro de América Latina [Internet]. Nov-11-2005. Available from: http://www.cronica.com.mx/notas/2005/212974.html [Accessed: Aug-12-2017]

[24] U.S. Geological Survey. Landsat 4-7 Climate Data Record (CDR) Surface Reflectance, Product Guide. USA: Department of the Interior U.S. Geological Survey; 2015. 26 p

[25] U.S. Geological Survey. Provisional Landsat 8 Surface Reflectance Product. Product Guide. Geological Survey, USA: Department of the Interior U.S; 2015. 27 p

[26] U.S. Geological Survey. What are the band designations for the Landsat satellites? [Internet]. 2017 [Updated: Jun-22-2017]. Available from: https://landsat.usgs.gov/what-are-band-designations-landsat-satellites [Accessed: 25-05-2017]

[27] Masek JG, Vermote EF, Saleous NE, Wolfe R, Hall FG, Huemmrich KF, Gao F, Kutler J, Lim TA. Landsat surface reflectance dataset for North America, 1990-2000. Geoscience and Remote Sensing Letters, IEEE. 2006;**3**:68-72. DOI: 10.1109/LGRS.2005.857030

[28] Lu, D., Moran, E., Hetrick, S., Li, G. Land-use and land-cover change detection. In: Advances in Environmental Remote Sensing Sensors, Algorithms, and Applications. New York: Press Taylor & Francis Group; 2011. p. 273-290

[29] Chen X, Li J, Zhang Y, Tao L. Change detection with multi-source defective remote sensing images based on evidential fusion. In: Commission VII, WG VII/5, editors. XXIII ISPRS Congress, Remote Sensing and Spatial Information Sciences; 12-19-07-2016; Prague, Czech Republic. Copernicus Publications; 2016. p. 125-132. DOI: 10.5194/isprsannals-III-7-125-2016

[30] Soenen SA, Peddle DR, Coburn CAA. Modified sun-canopy-sensor topographic correction in forested terrain. IEEE Transactions on Geoscience and Remote Sensing. 2005;**43**(9):2148-2159. DOI: 10.1109/TGRS.2005.852480

[31] Gu D, Gillespie A. Topographic normalization of Landsat TM images of forest based on subpixel sun-canopy-sensor geometry. Remote Sensing of Environment. 1998;**64**(2):166-175

[32] Teillet PM, Guindon B, Goodenough DG. On the slope-aspect correction of multispectral scanner data. Canadian Journal of Remote Sensing. 1982;**8**(2):84-106. DOI: 10.1080/07038992.1982.10855028

[33] Vázquez-Jiménez R, Romero-Calcerrada R, Ramos-Bernal RN, Arrogante-Funes P, Novillo CJ. Topographic correction to landsat imagery through slope classification by applying the SCS + C method in mountainous forest areas. International Journal of Geo-Information. 2017;**6**(287):1-16. DOI: 10.3390/ijgi6090287

[34] Kauth RJ, Thomas G. The tasselled cap. A graphic description of the spectral-temporal development of agricultural crops as seen by landsat. In: Symposium on Machine Processing of Remotely Sensed Data; 29 Jun-01 Jul. 1976; West Lafayette Indiana. USA. IEEE; 1976. p. 41-51

[35] Crist EP. A TM Tasseled Cap equivalent Transformation for Reflectance Factor Data. Remote Sensing of Environment. 1985;**17**(3):301-306

[36] Baig MHA, Zhang L, Shuai T, Tong Q. Derivation of a tasselled cap transformation based on Landsat 8 at-satellite reflectance. Remote Sensing Letters. 2014;**5**(5):423-431. DOI: 10.1080/2150704X.2014.915434

[37] Hotelling H. Analysis of a complex of statistical variables into principal components. Journal of Educational psychology. 1933;**24**(6):417-441. DOI: 10.1037/h0071325

[38] Lu D, Mausel P, Brondízio E, Moran E. Change detection techniques. International Journal of Remote Sensing. 2004;**25**(12):2365-2401. DOI: 10.1080/0143116031000139863

[39] Mahalanobis PC. On the generalized distance in statistics. Proceedings of the National Institute of Sciences (Calcutta). 1936;**12**:49-55

[40] McLachlan GJ. Mahalanobis distance. Resonance. 1999;**4**(6):20-26

[41] Wicklin, R. What is Mahalanobis distance?. The DO Loop. Statistical programming in SAS with an emphasis on SAS/IML programs [Internet]. 2012 [Updated: Feb-15-2012]. Available from: http://blogs.sas.com/content/iml/2012/02/15/what-is-Mahalanobis-distance.html [Accessed: Aug-31-2017]

[42] Reimann C, Filzmoser P, Garrett R, Dutter R. Statistical Data Analysis Explained: Applied Environmental Statistics with R. England: John Wiley & Sons; 2011. 362 p

[43] Cohen J. A coefficient of agreement for nominal scales. Educational and Psychological Measurement. 1960;**20**:37-40

[44] Couto P. Assessing the accuracy of spatial simulation models. Ecological Modelling. 2003;**167**(1-2):181-198. DOI: 10.1016/S0304-3800(03)00176-5

Chapter 10

Clouds Motion Estimation from Ground-Based Sky Camera and Satellite Images

Ali Youssef Zaher and Afraa Ghanem

Additional information is available at the end of the chapter

http://dx.doi.org/10.5772/intechopen.71263

Abstract

Estimation of cloud motion is a challenging task due to the non-linear phenomena of cloud formation and deformation. Satellite images processing is a popular tool used to study the characteristics of clouds which constitute major factors in forecasting the meteorological parameters. Due to the low resolution of satellite images, researchers have turned towards analyzing the high-resolution images captured by ground-based sky cameras. The first objective of this chapter is to demonstrate the different techniques used to estimate clouds motion and to compare them with respect to the accuracy and the computational time. The second aim is to propose a fast and efficient block matching technique based on combining the two types of images. The first idea of our approach is to analyze the low-resolution satellite images to detect the direction of motion. Then, the direction is used to orient the search process to estimate the optimal motion vectors from the high-resolution ground-based sky images. The second idea of our method is to use the entropy technique to find the optimal block sizes. The third idea is to imply an adaptive cost function to perform the matching process. The comparative study demonstrates the high performance of the proposed method with regards to the robustness, the accuracy and the computation time.

Keywords: block matching, exhaust search, three step search, four step search, optical flow, Lucas-Kanade, Horn-Schunck

1. Introduction

Clouds constitute major factors in determining the radiation budget of the Earth, and consequently, play crucial roles in modulating the climate for both local and global scales. Satellite images processing is an essential tool for both meteorologists and scientists to collect large-scale cloud information. Most of meteorological satellites are placed on geostationary orbit providing successive images in different spectral channels such as infrared, water vapor and visible channels [1]. Chiefly, the sequence resulting by collecting successive images have been useful to study the characteristics and the spatial distribution of clouds. Many satellite image based studies have focused on detecting the movement of clouds to forecast the solar irradiance which is essential for the effective operation of various solar applications such as solar thermal systems, photovoltaic systems, and grid regulation [2]. However, the low resolution of satellite images with respect to space and time is not adequate to satisfy the requirements of many real-time applications. Thus, researchers have used the images captured by ground-based cameras in order to make up the deficiency of satellite cloud observations [3].

In the context of motion estimation, block matching and optical flow are the most popular of all motion estimation techniques due to their effectiveness and simplicity. Optical flow methods are based on brightness constancy of matching pixels in consecutive images. Ideally, optical flow would be the same as the motion field, but the apparent motion can be caused by lighting changes without any motion. Thus, smoothness constraints are used by researchers to overcome the limitations of brightness constancy assumption [4].

Concerning block matching based approaches, the consecutive images are partitioned into macro blocks of pixels. And then, for each block at the current image, the algorithm seeks for the best matching block within a search area from the previous image according to a matching criterion. The exhaustive search ES provides the best performance by matching all possible blocks within the search area, but the search process takes a significant computation time [5]. Thus, fast algorithms have been proposed such as three step search 3SS, four step search 4SS, and diamond search DS algorithms [6].

In block based methods, block size has a large impact on the efficiency. All the approaches with fixed sized blocks have low performance because larger blocks may have some objects that move in different directions and at different speeds, on the other hand, smaller blocks are difficult to be recognized correctly [7].

Our first aim in this chapter is to provide an overview of the principal methods used for clouds motion estimation. The second purpose is to propose a fast and efficient block matching algorithm. The proposed algorithm has improved the estimation accuracy by using: two types of images, variable block sizes instead of fixed block sizes, and an adaptive cost function to perform the matching process.

Thus, this chapter is organized as follows: Section 2 describes the data analyzed in this study. In Sections 3 and 4, the popular block matching and optical flow approaches are presented. Section 5 is dedicated to the proposed algorithm. Section 6 shows the experimental results and a comparative study between the algorithms under different types of movements. Finally, Section 7 presents our conclusions.

2. Data collection

The development of our methodology for cloud motion detection has been performed on analyzing the images captured by METEOSAT second generation satellites MSG, and those taken by ground-based camera systems (**Figure 1**).

(a) (b)

Figure 1. Sample of analyzed images: (a) gray scale satellite image, and (b) true color all-sky image.

The MSG satellite series are operated by the European organization for the exploration of meteorological satellites (EUMETSAT). They provide image data in visible and infrared channels that cover Europe, Africa and the Atlantic Ocean with a baseline repeat cycle of 15 minutes. We are interested in the visible images of the channel 0.6 μm which have a normal resolution of 3×3 km^2.

In fact, the METEOSAT sensor for the visible channel captures sunlight of the visible band reflected by an object back to the satellite. The fraction of solar radiation that is reflected back to space is called the albedo. The cloudy surfaces usually have a higher albedo than the other surfaces of the Earth. So, clouds usually appear white, while land and water surfaces appear in darker gray scale levels.

The ground-based images are taken by diverse whole sky camera systems (WSC) distributed on different geographic locations in France, Italy and Switzerland. The data are acquired in several seasons; therefore, they cover a wide range of possible sky conditions.

The whole sky camera consists basically of a digital color video camera mounted with a fish-eye lens that forms circular fish-eye images with 180-degree field of view across the image circle. The time interval between two successive images varies from 15 s to 5 minutes according to the application requirements. Images have a resolution of 768×576 pixels, and are stored as JPEG images. In these images, when clouds are thin they appear as white because they scatter and reflect all the visible colors of light which are combined to give white light. If clouds are thick, they will absorb a significant amount of the sunlight, and consequently, they will appear grayish and darker.

3. Block-based motion estimation methods

Block matching algorithms for motion estimation play a key role in most of video and image processing applications such as objects tracking, video compression standards, medical image analysis, video telephony and real-time video conferencing, etc.

The underlying assumption behind block matching algorithms is that the visual contents of an object in a frame of video sequence are closely related to those of the same object on the subsequent frame. The main idea of these algorithms is to divide the two successive images into a matrix of macro blocks with size of $L \times M$, as presented in **Figure 2**. Then, each block of the current image is considered as a reference block. Afterwards, the reference block is compared to all blocks within a search window with size of $(2W + L) \times (2W + M)$ of the previous image to determine the block with a higher match to the reference block.

In the literature, various cost functions are used as similarity measures to match the reference macro block to the candidates blocks such as: mean absolute differences *MAD* [8], or sum of absolute differences *SAD* [9, 10]. These criteria are expressed by the following formulas:

$$MAD = \frac{1}{L \times M} \sum_{i=1}^{L} \sum_{j=1}^{M} \left| I_n(x_r + i, y_r + j) - I_{n-1}(x + i, y + j) \right| \quad (1)$$

$$SAD = \sum_{i=1}^{L} \sum_{j=1}^{M} \left| I_n(x_r + i, y_r + j) - I_{n-1}(x + i, y + j) \right| \quad (2)$$

Where:

- $I_n\,(x_r, y_r)$ is the reference block of the current frame (n) and (x_r, y_r) is the coordinate of its upper left corner.
- $I_{n-1}(x, y)$ is the candidate block of the previous frame (n–1) and (x, y) is the coordinate of its upper left corner.

Once the best matching block is located, the motion vectors *MVs* can be obtained by calculating the displacement between the block in the present image and the best matched block in the previous image in horizontal and vertical directions.

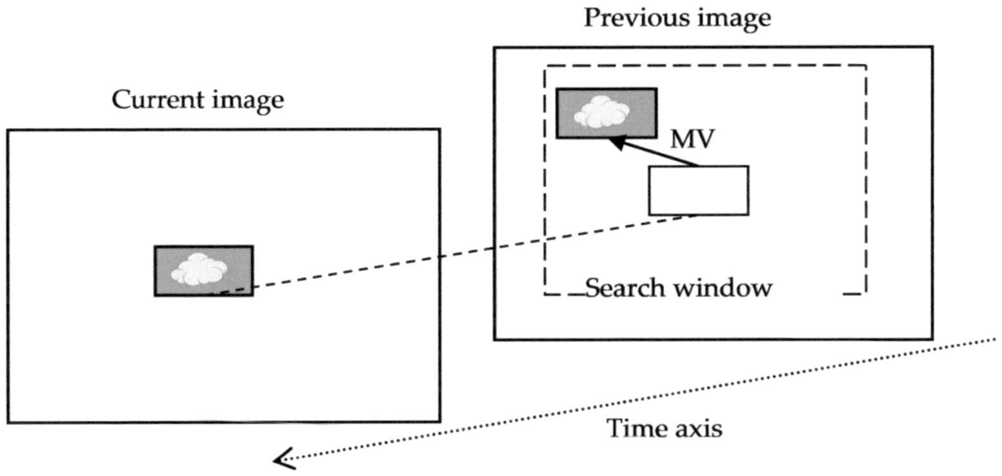

Figure 2. Block matching process.

The search-window size and the block size have a significant impact on the performance of block matching algorithms with regard to the accuracy and the computation time.

The search area size depends strongly on the object's velocity, if object's speed is high and small search windows are used, the block that matches best with the reference block will be outside the search region and the estimate will be unreliable. If the objects move at low speeds and the search window is too large, a large number of matching blocks can be found in the search zone and the estimate will be wrong. Further, the processing time is greatly increased.

Concerning the block size, the selection of this parameter is essential for any block-based motion estimation algorithm. If the blocks are too small, the number of matching blocks increases as well as the estimation error and the computation time. On other hand, if the blocks are very big, there is a high probability that the block contains different objects moving at different speeds, and consequently, the estimation error is increased as well.

In order to reduce the impacts of the above-mentioned parameters, dynamic search area and block sizes are preferred to find a reasonable balance between the processing time and the desired accuracy according the application.

3.1. Three step search approach

Three step search 3SS is one of the earliest methods used to optimize the search process with a performance close to that can be obtained from exhaustive search algorithm [11]. This method is based on three steps fine-coarse search mechanism. The basic idea is to conduct the research process in raster order. In other words, the block with minimum distortion in the previous step is used to determine the positions of the blocks that must be examined in the current step. **Figure 3** shows an example to illustrate the three-step search process for $W = 7$.

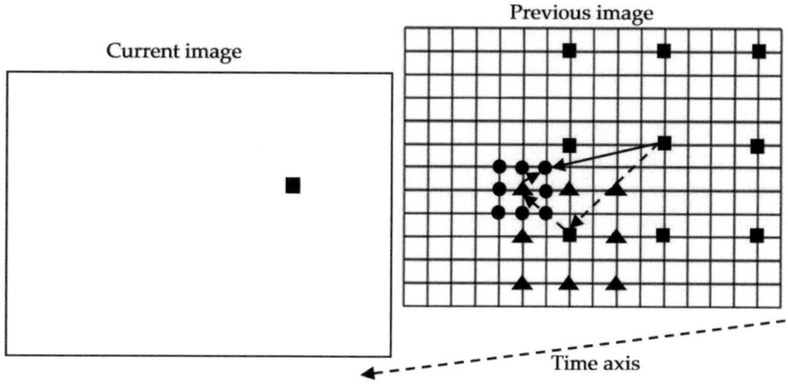

Figure 3. Three step search process with $W = 7$.

In the first step, eight blocks at a distance $d_1 = W/2$ from the center block (x_r, y_r) are determined in the previous image. Then the eight blocks in addition to the center block are compared to the reference block of the current image. The block with minimum distortion is considered as a center block in the next step. In the second step, a new eight blocks are determined at a distance $d_2 = d_1/2$ from the new center block. Then the matching operation is done to find the new center block as in the first step. In the third step, the distance from the center block becomes $d_3 = 1$ pixel and the best matched block resulted from this step is used to calculate the motion vectors.

As can be seen, the number of matches are significantly reduced to 27 operations comparing to 225 operations used in the exhaust search algorithm.

3.2. Four step search approach

The principle of four-step search algorithm is based on center biased searching process as well as in three-step search [12]. This method is composed of four steps. In the beginning, the size of the search window is fixed at W pixels. Afterwards, eight blocks around the center block are determined at a distance $d_1 = W/4$ pixels from the center block, and then, they are checked as in the three-step search process. The block with a higher match is kept to be the center in the next step unless it is the center of the search window then the search process jumps immediately to the fourth step. The following two steps are exactly the same as the first step. In the fourth step, the distance from the center block is a half of that used in the beginning $d_4 = d_1/2$, and the location of the best matched block is used to estimate the motion vectors of the reference block.

Contrary to three-step search algorithm, the number of matching operations to reach the best matched block is variable due to the condition imposed in the first three steps. With regard to the example presented in **Figure 4**, the number of checked blocks equals 36 blocks. This indicates that, the computation time of four-step search process is greater than that of three-step search, but it is still less than that of full search process.

3.3. Diamond search approach

As its name indicates, the diamond search method DS implies a diamond shaped search windows instead of square search windows which are used in the above-mentioned approaches [13]. In addition, the number of steps needed for the convergence is not limited.

Firstly, a large diamond-shaped window is used and eight blocks around the center block are examined. If the center of the search area has the best match then the procedure jumps to the last step, else the best matched block will be the center of the search window in the next step. The procedure is repeated until the match occurs with the center of the search window then, a smaller diamond-shaped search window is used and the block with the minimum distortion is used to determine the motion vectors.

The principle of diamond search process is presented in **Figure 5**. The number of examined blocks needed to reach the best match is equal to 9 + 9 + 9 + 9 blocks and this is equivalent to the number of operations used in four step search process.

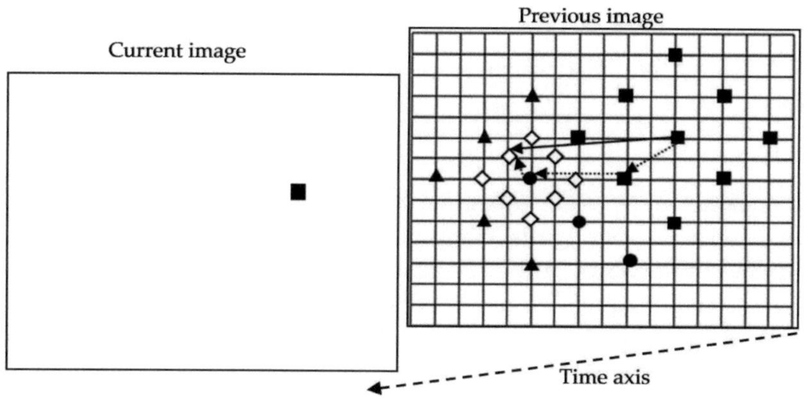

Figure 4. Four step search process.

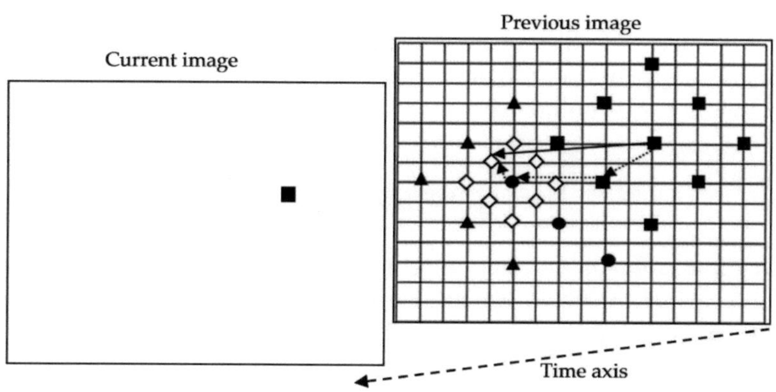

Figure 5. Diamond search process.

4. Optical flow methods

Block matching algorithms divide the consecutive images into blocks of pixels which move at the same speed and in the same direction. Whereas, optical flow algorithms are pixel-based approaches that determine motion vectors for every pixel in the image.

The last methods work on brightness constancy assumption, that is, two corresponding pixels in consecutive frames should have the same brightness. This assumption leads us to the optical flow equation [14, 15]:

$$\left(\frac{\partial I}{\partial x}\frac{\partial x}{\partial t} + \frac{\partial I}{\partial y}\frac{\partial y}{\partial t}\right) + \frac{\partial I}{\partial t} = \left(\frac{\partial I}{\partial x}u + \frac{\partial I}{\partial y}v\right) + \frac{\partial I}{\partial t} = I_x u + I_y v + I_t = 0 \quad (3)$$

Where:

- I_x, I_y and I_t are the partial derivatives of image brightness with respect to x, y and t respectively.
- u is the horizontal velocity and v the vertical velocity.

Since the last equation involves two unknown variables u and v, there are infinite solutions that satisfy its constraints. In other words, many pixels in the previous frame could match with a given pixel in the current frame. Thus, the problem of motion estimation is ill-posed. It is for this reason that various methods are proposed to reduce the number of solutions by introducing additional constraints on the smoothness of the motion vector field. In this context, the following two methods are among the most common methods that use regularization constraints.

4.1. Horn and Schunk approach

Horn and Schunck added an additional regularization condition that reduces the solutions space of the optical flow equation. They transformed the last equation into an optimization problem that minimizes both the optical flow constraint and the magnitude of the variations of the flow field. In this context, the flow optic is formulated as a global energy function E which is then sought to be minimized as follows [16]:

$$E = \iint \left(I_x u + I_y v + I_t\right)^2 dxdy + \alpha \iint \left(\left(\frac{\partial u}{x}\right)^2 + \left(\frac{\partial u}{y}\right)^2 + \left(\frac{\partial v}{x}\right)^2 + \left(\frac{\partial v}{y}\right)^2\right) dxdy \tag{4}$$

where: α is a regularization parameter which influences the smoothness of the motion vectors. It is usually adjusted empirically depending on the desired smoothness. Larger values of α give a smoother motion filed.

The minimization problem can be solved by using an iterative process through which we can estimate the new set of the motion vectors (u^{n+1}, v^{n+1}) from the estimated derivatives and the average of the previous velocity estimates (u^n, v^n) as follows:

$$u^{n+1} = u^{-n} - \frac{I_x\left(I_x u^{-n} + I_y v^{-n} + I_t\right)}{\alpha + I_x^2 + I_y^2} \tag{5}$$

$$v^{n+1} = v^{-n} - \frac{I_y\left(I_x u^{-n} + I_y v^{-n} + I_t\right)}{\alpha + I_x^2 + I_y^2} \tag{6}$$

Considering two consecutive images and a pixel of coordinates (i, j) the spatial and temporal derivatives I_x, I_y, I_t can be calculated using the following approximations:

$$I_x \approx \frac{1}{4}\left[I_{i,j+1,k} - I_{i,j,k} + I_{i+1,j+1,k} - I_{i+1,j,k} + I_{i,j+1,k+1} - I_{i,j,k+1} + I_{i+1,j+1,k+1} - I_{i+1,j,k+1}\right] \tag{7}$$

$$I_y \approx \frac{1}{4}\left[I_{i+1,j,k} - I_{i,j,k} + I_{i+1,j+1,k} - I_{i,j+1,k} + I_{i+1,j,k+1} - I_{i,j,k+1} + I_{i+1,j+1,k+1} - I_{i,j+1,k+1}\right] \tag{8}$$

$$I_t \approx \frac{1}{4}\left[I_{i,j,k+1} - I_{i,j,k} + I_{i+1,j,k+1} - I_{i+1,j,k} + I_{i,j+1,k+1} - I_{i,j+1,k} + I_{i+1,j+1,k+1} - I_{i+1,j+1,k}\right] \tag{9}$$

4.2. Lucas and Kanade approach

Horn-Schunck approach gives an appropriate solution for the optical flow equation, but its iterative process is computationally expensive. Lucas and Kanade have overcome this problem by implying the least square method that works over an area of neighboring pixels to find the best matching pixels. This method assumes that the displacement of visual contents between two successive frames is small and approximately constant within a neighborhood of the pixel under consideration. Thus, the current image is divided into smaller zones of pixels in each of them the motion vectors are considered constant. Then, the basic optical flow equations for all pixels in a given zone Ω can be expressed mathematically as follows [17]:

$$\begin{aligned} I_x(x_1, y_1)u + I_y(x_1, y_1)v &= -I_t(x_1, y_1) \\ I_x(x_2, y_2)u + I_y(x_2, y_2)v &= -I_t(x_2, y_2) \\ &\vdots \\ I_x(x_n, y_n)u + I_y(x_n, y_n)v &= -I_t(x_n, y_n) \end{aligned} \tag{10}$$

Where x_n and y_n are pixel coordinates inside the zone Ω.

These equations can be written in the matricial form as:

$$\underbrace{\begin{bmatrix} I_x(x_1, y_1) & I_y(x_1, y_1) \\ I_x(x_2, y_2) & I_y(x_2, y_2) \\ \vdots & \vdots \\ I_x(x_n, y_n) & I_y(x_n, y_n) \end{bmatrix}}_{A} \underbrace{\begin{bmatrix} u \\ v \end{bmatrix}}_{V} = -\underbrace{\begin{bmatrix} I_t(x_1, y_1) \\ I_t(x_2, y_2) \\ \vdots \\ I_t(x_n, y_n) \end{bmatrix}}_{B} \tag{11}$$

Consequently, the motion vectors can be expressed using the following equation:

$$V = (A^T A)^{-1} A^T B \tag{12}$$

The last equation gives the same weight to all pixels in the zone Ω. In fact, it is better to give more weight to the pixels close to the center of the zone Ω.

5. Proposed algorithm

The developed approach is a new block matching algorithm with oriented search process. This approach is based on two principal parts: the aim of first part is to estimate the direction of cloud motion using the low-resolution satellite imagery. In the second part, we introduce the obtained direction into the search process of a variable size block matching algorithm that must be applied to the high-resolution ground-based images.

In fact, detecting the direction of motion using satellite images is low expensive in terms of computation time. In other hand, the number of computations is also reduced by adopting an oriented search process in lieu of full search process in the second part. In addition, the high-resolution images and the use of variable block size play an important role to increase the accuracy.

Our method can be summarized by the following steps:

Step 1. Detection of motion direction:

To detect the direction of movement of cloudy objects, we determine the zone of interest on the satellite images. Then, the exhaust search block matching algorithm is applied to the two consecutive satellite images to obtain the motion vectors u and v. Finally, the direction of movement θ is computed by the following formula:

$$\theta = \tan^{-1}\left(\tfrac{u}{v}\right) \tag{13}$$

Step 2. Determining the block size

As mentioned in Section 3, the block must be large enough to contain pixels with varied details for reducing the number of matches in the search window. In other words, the block must be textured.

We have used the entropy of a given block to characterize the texture of this block. If the entropy is large enough, the block is considered as textured block.

To apply the entropy method, firstly, the image is divided into small blocks and the entropy En is calculated for each block as follows [18, 19]:

$$En = -\sum_{i=1}^{N} P_i \, log_2(P_i) \tag{14}$$

Where: N is the number of possible gray levels, and P_i is a scalar contains the histogram count for the gray level i.

If the block has an entropy larger than a threshold then it is considered as textured block otherwise the block is considered as a smooth block.

After classifying the blocks, the number of textured blocks in the whole image is compared to a threshold value if this number is bigger than the threshold then the block size can be accepted in the block matching algorithm, else, the block size must be increased and all the operations will be repeated until obtaining the optimal block size.

Step 3. Application of the oriented search process

In this step, the two consecutive images I_n and I_{n-1} are divided into macro blocks with the optimal size obtained in the previous step. Then, each reference block (x_r, y_r) in the image I_n is compared to the blocks (x, y) of the image I_{n-1} in the motion direction θ. The coordinates (x, y) of each examined block are given by:

$$x = x_r + w \times \sin\theta \tag{15}$$

$$y = y_r + w \times \sin\theta \tag{16}$$

Where w takes the values from 0 to W which is the size of the search window.

The criterion used to match the candidates to the reference block in the search window is a cost function of the sum of absolute differences SAD and the Euclidean distance D between the reference block and the block candidate. This function is given by:

$$Cost = (1-\alpha)SAD + \alpha D \tag{17}$$

The parameter α is a weighting parameter and takes values between 0 and 1. For the low displacements α is high and vice versa. The best match occurs when the cost function $Cost$ is minimum. The use of the distance in the cost function limits the number of matches in the search window and this minimizes the error in estimating the motion vectors.

6. Results and discussion

6.1. Data set description

The database used for implementation and validation of our method is made of two types of images:

- High resolution RGB images (HR) collected every minute of daylight by whole sky imager. Each of them consists of 768 × 576 pixels with 8 bits per channel.
- Low resolution METEOSAT images in the visible channel acquired at 15 minutes intervals. These images are 8-bit color images with a resolution of 745 × 1054 pixels. For each of them, only 600 × 700 pixels representing Europe are considered.

To cover the representative conditions of different types of sky conditions, we have screened the complete data acquired during the period from September 2016 to May 2017 and we have selected approximately 1500 whole sky images and 1500 satellite images taken at the same time. The selection procedure depends upon clouds fraction thresholds. In this procedure, images with cloud fraction lower than 10% are considered as clear sky images and they are discarded from the data base because they do not provide meaningful motion fields. Furthermore, the images representing overcast conditions (cloud fraction ≥ 95%) usually have texture-less cloudy objects and the error in the estimation is very big for all algorithms; therefore, this type of images is not included in our database. 300 images (20% of the data base) are used as training set to optimize the parameters of algorithms such as block size for block matching algorithms and the smoothness area and the number of iterations for optical flow based algorithms. The other images of the database (1300 images) are used in the test stage to show the performances of the implemented algorithms.

6.2. Experimental methodology

In this work, the experimental methodology includes two stages: the training stage and the test stage. In the training stage, 300 images (20% of the data base) are used as training set to optimize the parameters of algorithms such as block size for block matching algorithms and the smoothness area for optical flow based algorithms. The other images of the database (1300 images) are used in the test stage to show the performances of the implemented algorithms.

In order to select the optimal parameters for each algorithm, we have estimated the motion vectors using the images in the training data set for different values of the parameters and then, we have kept the parameters that provide the best accuracy.

The accuracy is characterized by the *PSNR* value (Peak-Signal-to-Noise-Ratio). Such parameter quantifies the differences between an original image and its distorted version that is reconstructed using the estimated motion vectors. The bigger values of *PSNR* indicate a better accuracy of the algorithm. The *PSNR* is defined as [20]:

$$PSNR = 10\, log_{10}\left(\frac{I_{max}^{\ 2}}{MSE}\right) \tag{18}$$

Where, I_{max} is the maximum brightness and *MSE* is the mean square error between the original image and the reconstructed one.

The parameters of the block matching algorithm are selected experimentally by applying different values for each parameter and keeping those giving the better *PSNR*. Considering the block size, the best *PSNR* is obtained for a block size of 8 × 8 pixels as presented in **Figure 6**. Concerning the search area size, it depends on the maximum clouds displacement speed. For our experiments, the maximum displacement speed is 16 pixels, thus, the search range *W* is limited to 16 pixels.

For optical flow algorithms, the smoothness window Ω is fixed at 8 × 8 pixels as the block size in block matching algorithms.

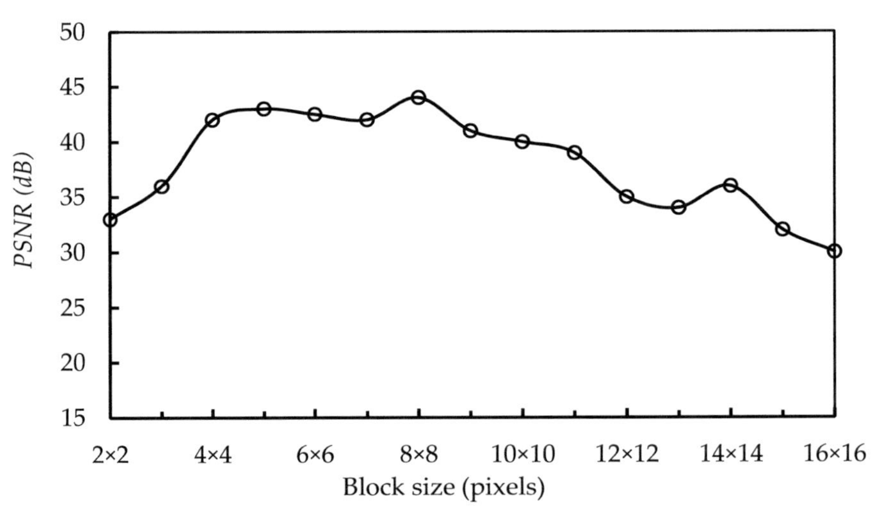

Figure 6. PSNR vs. block sizes.

6.3. Experimental results

All the above conventional algorithms have been implemented to estimate the displacement of clouds using sequences of high-resolution images captured by ground-based sky cameras. For our method, we have used the low-resolution METEOSAT satellite imagery and the high-resolution ground based images. Such sequences consist of different degrees and types of motion. Each image is divided into macro-blocks with the sizes of 8 × 8 pixels for all algorithms with fixed block size. But for our algorithm, the block size is variable and it takes values from varies from 4 × 4 to 12 × 12 pixels. The maximum displacement between two consecutive images is 16 × 16 pixels, thus, the optimal search window size W is ±16 pixels in both horizontal and vertical directions.

For evaluating the implemented algorithms, they are compared with regards to their accuracy and the computation time.

The accuracy is evaluated using the *PSNR* value. In addition to the accuracy, the standard deviation of the *PSNR* is calculated to qualify the robustness of the algorithms. A low standard deviation indicates that the *PSNR* values are close to the mean of the set and consequently the robustness is very high, while a high standard deviation indicates that the *PSNR* is spread out over a wide range of values and this makes the algorithm non-robust.

The *PSNR* comparison of the compensated images generated using the implemented algorithms is presented in **Figures 7** and **8**. **Table 1** indicates data for the standard deviation of the *PSNR* and the average computation time.

From these results, it is observed that the proposed method provided the maximum values of the *PSNR* for all types of motion and the minimum value of standard deviation, which means that our method has the best performance in terms of accuracy and robustness. On the other

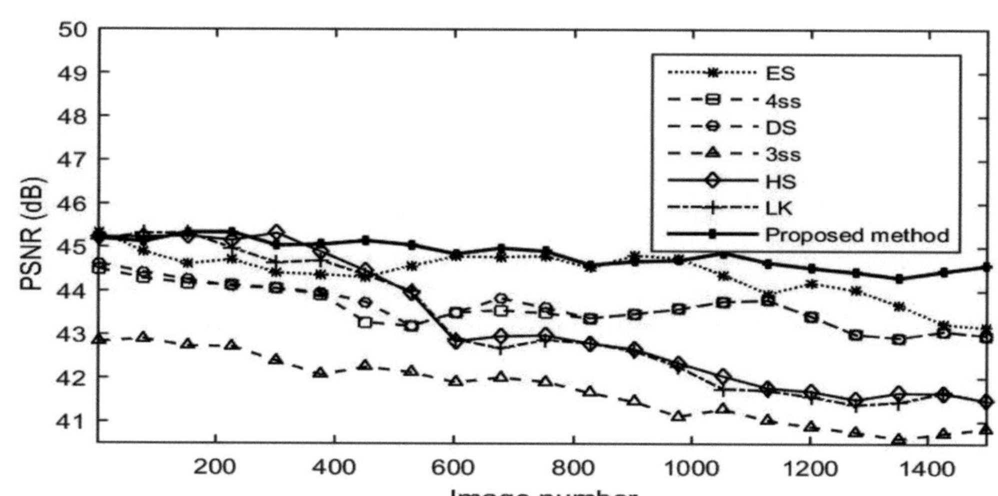

Figure 7. *PSNR* performance evaluation.

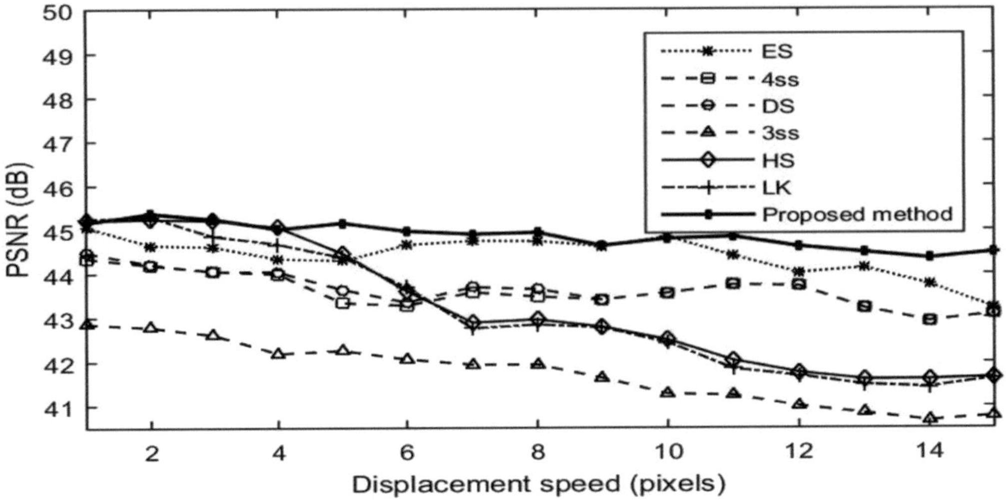

Figure 8. *PSNR* versus displacement speed.

Method	Average *PSNR* (dB)	Standard deviation	Computation time (s)
ES	44.3	0.49	13.2
3SS	41.7	0.73	1.8
4SS	43.6	0.41	2.5
DS	43.7	0.43	3.2
HS	43.22	1.41	10.4
LK	43.12	1.4	7.4
Proposed method	44.9	0.3	[4–6.8]

Table 1. Standard deviation and average values of computation time for all implemented algorithms.

hand, the computation time of our algorithm is highly decreased compared to the computation time of exhaust search algorithm and this is due to the use of an oriented search approach.

For the other algorithms, the exhaust search block matching method presents the better accuracy but it is the most expensive with regards to the computational time, and that is because of the great number of matching operations which equals to $(2W+1) \times (2W+1)$.

Furthermore, optical flow algorithms have a good accuracy for the low motions but this accuracy decreases significantly at high speeds. This is because moving objects do not preserve their intensity values from image to image in case of high motions.

The three-step search algorithm has the best computation time, but it also has the worst *PSNR* performance.

7. Conclusion

This chapter provides an overview of the popular techniques used for clouds motion estimation. In addition, it presents a fast and efficient block matching algorithm based on combining low-resolution satellite images and high-resolution ground based sky images.

In our algorithm, the use of an adaptive cost function and dynamic block sizes allowed us to obtain an accuracy better than that obtained by the conventional methods that treat one type of images and fixed block sizes. Furthermore, the oriented search process reduced the computation time of the proposed algorithm considerably comparing to the exhaust search and optical flow algorithms.

Author details

Ali Youssef Zaher[1]* and Afraa Ghanem[2]

*Address all correspondence to: ali.zaher@promes.cnrs.fr

1 University of Perpignan, PROMES-CNRS Laboratory, Perpignan, France

2 University of Perpignan, Perpignan, France

References

[1] Linares-Rodriguez A, Ruiz-Arias JA, Pozo-Vazquez D, Tovar-Pescador J. An artificial neural network ensemble model for estimating global solar radiation from Meteosat satellite images. Energy. 2013;**61**:636-645

[2] Alonso J, Ternero A, Batlles FJ, López G, Rodríguez J, Burgaleta JI. Prediction of cloudiness in short time periods using techniques of remote sensing and image processing. Energy Procedia. 2014;**49**:2280-2289

[3] Marquez R, Coimbra CFM. Intra-hour DNI forecasting based on cloud tracking image analysis. Solar Energy. 2013;**91**:327-336

[4] Chen W, Mied RP. Optical flow estimation for motion-compensated compression. Image and Vision Computing. 2013;**31**(31):275-289

[5] Liaw YC, Lai JZ, Hong ZC. Fast block matching using prediction and rejection criteria. Signal Processing. 2009;**89**(6):1115-1120

[6] Cuevas E, Zaldívar D, Pérez-Cisneros M, Sossa H, Osuna V. Block matching algorithm for motion estimation based on Artificial Bee Colony (ABC). Applied Soft Computing. 2013;**13**(6):3047-3059

[7] Khan NA, Masud S, Ahmad A. A variable block size motion estimation algorithm for real-time H. 264 video encoding. Signal Processing: Image Communication. 2006; **21**(4):306-315

[8] Pandian SIA, Bala GJ, Anitha J. A pattern based PSO approach for block matching in motion estimation. Engineering Applications of Artificial Intelligence. 2013;**26**(8):1811-1817

[9] Saha A, Mukherjee J, Sural S. A neighborhood elimination approach for block matching in motion estimation. Signal Processing: Image Communication. 2011;**26**(8):438-454

[10] Basha SM, Kannan M. Design and implementation of low-power motion estimation based on modified full-search block motion estimation. Journal of Computational Science. 2017;**21**:327-332

[11] Kulkarni SM, Bormane DS, Nalbalwar SL. Coding of video sequences using three step search algorithm. Procedia Computer Science. 2015;**49**:42-49

[12] Po LM, Ma WC. A novel four-step search algorithm for fast block motion estimation. IEEE Transactions on Circuits and Systems for Video Technology. 1996;**6**(3):313-317

[13] Zhu S, Ma K. A new diamond search algorithm for fast block-matching motion estimation. IEEE Transactions on Image Processing. 2000;**9**(2):287-290

[14] Wei SG, Yang L, Chen Z, Liu ZF. Motion detection based on optical flow and self-adaptive threshold segmentation. Procedia Engineering. 2011;**15**:3471-3476

[15] Sengar SS, Mukhopadhyay S. Motion detection using block based bi-directional optical flow method. Journal of Visual Communication and Image Representation. 2017;**49**: 89-103

[16] Horn BKP, Schunck BG. Determining optical flow. Artificial Intelligence. 1981;**17**(1-3): 185-203

[17] Lucas B, Kanade T. An iterative image registration technique with an application to stereo vision. In: Seventh International Joint Conference on Artificial Intelligence; 1981; Vancouver, Canada; 1981. pp. 674-679

[18] Kim SE, Jeon JJ, Eom IK. Image contrast enhancement using entropy scaling in wavelet domain. Signal Processing. 2016;**127**:1-11

[19] Wu H, Miao Z, Wang Y, Chen J, Ma C, Zhou T. Image completion with multi-image based on entropy reduction. Neurocomputing. 2015;**159**:157-171

[20] Jalloul MK, Al-Alaoui MA. A novel cooperative motion estimation algorithm based on particle swarm optimization and its multicore implementation. Signal Processing: Image Communication. 2015;**39**:121-140